THE TRUTH ABOUT BEING A

WOMAN
in
CONSTRUCTION

THE TRUTH ABOUT BEING A

WOMAN in CONSTRUCTION

FINDING LIMITLESS

JESSICA LYNN, MBA

THE TRUTH ABOUT BEING A WOMAN IN CONSTRUCTION
Finding Limitless

Copyright © 2024
by Jessica Lynn, MBA

All rights reserved. No part of this publication may be reproduced, stored, or transmitted in any form or by any means, electronic, mechanical, photocopying, recording, scanning, or otherwise without written permission from the publisher. It is illegal to copy this book, post it to a website, or distribute it by any other means without permission.

Jessica Lynn, MBA, asserts the moral right to be identified as the author of this work.

Designations used by companies to distinguish their products are often claimed as trademarks. All brand names and product names used in this book and on its cover are trade names, service marks, trademarks, and registered trademarks of their respective owners. The publishers and the book are not associated with any product or vendor mentioned in this book.

None of the companies referenced within the book have endorsed the book. Names of individuals and companies have been changed to protect their privacy.

First edition
Advisor: Ambria Quick
Advisor: Amanda Luchetti

ISBN (paperback): 978-1-962280-61-7
ISBN (ebook): 978-1-962280-60-0

CONTENTS

Foreword ix
Preface xv

PART I 1
1. Start 3
   High School 4
   Feeling Lost 9
2. Welder 19
   Sheet Metal Shop 22
   Graduation 28
   Spark Of Light 33
   Refresh And Pivot 37
3. Construction Management 43
   Internship 51
   Summer Journey 56
   Graduation 58
4. California 65
   Refinery 74
5. Denver 87
   Dry Promotion 90
6. Oregon 99
   Structural Steel 107
7. Pandemic 121
   Bad Touch 126
   Problem Child 130
   Déjà Vu 136

PART II 147
8. MBA 149
   - Gaining Experience 152
   - Lean Practice 158
9. Austin 165
10. Pivot 181
    - Pivotal Moment 186
11. Director 193
    - Pivot Higher 204
12. Closing 207

Afterword 211
Vision for the Future 213
The Skilled Project 217
Pivot Higher + (Hireher) 219
References 223
About the Author 231

THANK YOU

To my mom, who put up with my rebellious streak and hard-headedness. Thank you for being there with me even now as I continue to map out my path. Your sacrifices in raising us three kids are what enabled me to be who I am today.

Ambria, my sister from another mister. Field sister and sister for life, we stay bonded regardless of what life throws our way. I love you dearly and am so grateful for your support and strength. You amaze me.

The two Dans, who were part of my story—Dan Johnson (R.I.P.) and Dan Hammond. You both were shining lights in some difficult periods of my life. Both of you showed me what it meant to be a caring and gentle man in this hard world. Dan Hammond, you believed in my ability and saw value in my insights before many others did (and still don't see). Thank you for being such an amazing human and caring person.

Tonya, Jayvetta, Tyler, Elynn, and Sylvain—you all were *lifesavers* during such a dark and turbulent time. My story is not mine alone but ours. You all kept it real with kindness while offering a safe space when others did not.

Elizabeth. I am beyond thankful for your support and leadership. In the brief time we had together, I learned so much.

The world needs more of you in it. You are a bright light leading the way for humanity through empathy.

Becky, I am so appreciative of your sponsorship and guidance. I was feeling stuck and unsure when you came in and connected me to the right people. You saw my brilliance before I did and helped guide me to those who would value it most.

Pivot Higher Plus Squad. *You women are amazing!* I am so humbled by your range of skills and abilities, as well as your commitment to making this world a better place. Thank you for trusting my vision and bringing your wealth of expertise into the mix to elevate what we do, together.

Last, to my amazing book team Mikey and Carly. I cannot express how much I appreciate your patience with all my extra S's and not enough commas. You helped bring my story to life in such a polished and thoughtful way.

FOREWORD

If I could sum up what it is like to be a woman in construction, I would say that the journey is hostile and vicious. It is no crystal staircase, yet it is exhilarating, worthwhile, and an experience like no other. Despite the obstacles I have gone through as a woman in construction, there is nothing more rewarding than seeing engineering drawings come to life in the fabrication shops and construction field-framing areas and knowing I had a hand in all of it.

I will never forget my first couple of weeks working as an office and field structural engineer. I had to weld, bevel, and hand-saw steel pipe and angle iron. And I used a torch to cut through steel plates. We had to go through this training to better understand the work we would be managing, scheduling, and tracking while fabricating and building offshore platforms. Honestly, while going through this training, I questioned why I even went to college to be an engineer instead of going to trade school. I loved welding! Unfortunately, when you are a young female growing up in a rural town in North Carolina, the only careers you are exposed to are education, manufacturing, and health care. Despite this, my transition from high school to

THE TRUTH ABOUT BEING A WOMAN IN CONSTRUCTION

college thankfully led me to discover the world of engineering and construction.

My journey through the construction industry was far from smooth sailing. I was faced with resistance—a ton of it. After all, I am Black, and I am a woman. Unfortunately, in this world, those are two strikes that people use to discriminate and count someone out in both the professional and personal realms. I had to overcome many obstacles while growing up in a world that was much less inclusive than it is today. Working in a male-dominated industry that is also mostly White, I had to constantly prove that I was intelligent, had a strong work ethic, and could work autonomously in a complex work environment. In other words, I was constantly battling the ignorant stereotypes surrounding being Black and being a woman. I worked myself to the point of burnout many times just to prove I was worthy of being in this industry. Meanwhile, I had male colleagues who would do the bare minimum and be praised and recognized for doing exactly that—the minimum.

Being a woman in construction can be isolating, lonely, and uneasy. There are times when my voice isn't heard, my presence isn't welcomed, and my ideas are not taken into consideration. Yet, I would see my male colleagues being met with welcoming arms. They make long-lasting friends and are invited out to lunch often. Management will even invite them on weekend fishing trips on their boats. As a Black woman, I was not extended the same invitations for lunch, my male colleagues refused to shake my hand, and management would not give me the support that was so readily available to my male colleagues even though my projects required those same resources.

Once, a colleague came to work drunk and tried to physically fight me and another coworker. What did HR do? Gave them a slap on the wrist. No suspension. Nothing. On another occasion, following a meeting in one of the construction trailers, an older man went on and on to a female colleague and me about how a woman's place was in the kitchen and in the bedroom. I still chuckle at that one because my female coworker and I were the hardest-working engineers on the project.

On that same project, our male counterparts failed to follow up on shop work orders, failed to track daily construction progress, and failed to keep the project on schedule. So, my female colleague and I picked up their slack because we actually cared about the project's success and wanted to make sure our work, and theirs, got done. We were the two main engineers keeping that project afloat. Yet that man had the nerve to tell us we didn't belong simply because we were women.

Thankfully, I have also had some pleasant times while working in construction. I had the pleasure of working with colleagues who were genuinely good people. They encouraged me and mentored me. They wanted to see me win even while so many wanted to see me fail. In particular, I'll never forget the construction structural superintendent, and field-framing superintendent I was assigned to on my first construction project. We were fabricating and constructing an offshore platform that now sits off the coast of Brazil. These three men saw my work ethic, acknowledged it, and entrusted me with great responsibility. When I had questions or fires, I needed to put out, they made sure I was supported. They knew that setting me up with the tools and resources for success meant success for the overall project. They didn't dwell on and become bothered by the fact that I was Black and female. They saw that I was willing to work, and they respected and supported that above all else.

I am so grateful for the colleagues who took me under their wings, mentored me, celebrated my accomplishments, and gave me constructive feedback so I could better myself in this industry. Yet, sadly, there is still much work to be done to make the world of construction more inclusive for women and people of color. This is why I am thankful for those along my journey who welcomed me and continue to work tirelessly to promote diversity and inclusion—especially colleagues like Jessi.

I met Jessi during a week-long work trip to Denver, Colorado. I had been selected, along with some other colleagues from my district, to visit the company's training facility. Thank goodness I accepted the offer. Had I not gone, I would have never met Jessi. She was from another district within the company and one of the few other women I met there. Little did I know that I would be meeting a warrior, a trailblazer, and a best friend. Jessi has become an essential part of my life, not only on a professional level but also on a personal one. She is my dear sister, my hype woman, my advocate, and someone I can vent to about the challenges we women face in the engineering and construction industry.

FOREWORD

I will forever admire her unwavering tenacity to fight for equality on every frontier, as well as her efforts to educate women on the careers available in construction. Her journey, as you will see in this book, is remarkable. She shows us that being a woman in construction isn't glamorous or for the faint of heart. It is no easy feat, but it is rewarding, and it is worth it. Not only are we setting ourselves up to be financially independent, but we are also honoring the women who came before us and showed us that we can hold our own in this industry. As for the women who will come into this field after us, I hope they will be inspired to enter construction and become leaders who will create a more inclusive and sustainable world.

This book is a piece of literature that is much needed in this day and age. It is one woman's journey that so many women, including me, can relate to and use as a guide as they navigate all male-dominated career fields. Jessi has laid out her journey, which consists of a bittersweet mixture of failures and accomplishments, weaknesses and strengths, and tough losses and wonderful wins. May this piece of literature inspire you, encourage you, push you, and motivate you to press forward amid the challenges we face as women in construction.

— *Ambria Quick*

PREFACE

Growing up as a child in the Midwest, I always assumed working hard was a normal way of life. Physical labor, sweaty summers, and learning how to do the work most adults wouldn't touch if they could choose was normal for my siblings and me. As a little girl working on the weekends with my father, who was a residential contractor in Indiana, could I have ever imagined I would grow into a woman with such a breadth of perspective and understanding of the male-dominated workforce? My hope is that this book inspires more young women to know that we can accomplish anything we set our minds to—but also that they must be aware of the difficulties ahead.

My story reflects experiences my female colleagues, friends, and I have shared, representing an entire industry in desperate need of hard redirection.

THE TRUTH ABOUT BEING A WOMAN IN CONSTRUCTION

Little Jessi Building

PART I

Part 1

CHAPTER 1

START

*B*lood, sweat, and tears have been threaded throughout my life. My parents had a landscaping business in conjunction with my father's residential construction business. As far back as I can remember, my two siblings and I spent many of our summers and weekends picking up branches, using a puck-shaped magnet to search for dropped nails, carrying shingles up and down a ladder, hanging drywall, and installing tile.

From an early point, it was instilled in us that everyone had to work. We were a very poor family living on the south side of Indy (Indianapolis), with little to no money to make ends meet. My mom would take a calculator to the store to ensure we didn't surpass our tiny meal budget. I remember the trips we had to take on the bus to go downtown to stand in line for food stamps.

Needless to say, this was what was familiar to me and mine. Once my mom was able to make changes in her career and start working as a substitute teacher, our lives migrated in a healthier, more wholesome direction.

When I was twelve, my mother divorced my father, and we **moved closer to Garfield Park. At the time, this was a poorer** part of the city but a step up from Orange Street, where my mom's clay pots would get shattered by aggressive teenagers in the neighborhood, and we would regularly hear gunshots at night. The work we kids performed didn't change too much. My father had us every other weekend, and he consistently fed us a diet of McDonald's and put us to work. All three of us **kids were always charged with doing backbreaking work.** The majority of the time, we stayed in line with our emotionally abusive father. If we didn't do something correctly or if we made a single complaint, there was a butt-whooping or an endless tirade of his belittling speeches involved.

Every weekend, we all learned new skills in operating **different types of mowing equipment, tools, and automotive** elements. By age fourteen, I was able to hook up and unload a **trailer, change out a weed eater line, and identify the differences** in paints and the surfaces they were associated with. I was stronger than the average teenager since many of the jobs my father took us on involved many trips to and from Lowes or Menards for plywood, drywall, or roofing shingles.

HIGH SCHOOL

High school is a confusing and difficult time. Discovering ourselves and the path to take seems to be all-consuming.

START

During this pivotal time in our formative years, especially as young women, it can be challenging to know what is going on inside of us and how to grow into adults.

For most teenagers, it is a terrifying time. Hormones are all over the place, the world is expanding, and ideas are forming. I was no different, except that I was going down a path that was leading to serious trouble.

After "breaking up" with my dad and creating a safe boundary for the first time in my life, I started to ping-pong all over the place and struggled to stay academically engaged with school. Most of the time, I would do the bare minimum or skip class. I enjoyed the freedom of being outside the school's jail-like walls and roaming around the neighborhood or nearby creek. It gave me a place to breathe and think.

Girl Band Photo

Emotionally, I was hurting after years of physical and mental abuse from my father. I was feeling insecure about my body as I started to blossom into a woman. My male role model had always made fun of me for wearing pink, and he expected me to more or less act like a boy. I was rough and tumble by nature, so the term *tomboy* wasn't a stretch. But anytime I wanted to explore something beyond skateboarding or football, my father criticized me.

I was the eldest of three, with my sister being two years younger and my brother being the baby. We did many things together and protected each other as best as we could. However, we all struggled to please our father and navigate his irrational and sporadic authority.

Then there was the physical aspect of my body as it changed in different ways, with my chest going from flat to a small A cup and periods driving a slew of hormones each month.

I got my first period in seventh grade over one of the weekends I was visiting my dad. These terrible pains in my abdomen kept growing, and I had no idea why. I had a volleyball game that weekend, and at some point on the way to the match, I started bleeding and didn't take note until partway through the first game. I went to the bathroom, and there it was—blood all over my blue net shorts. In one swooping moment, I was terrified and mortified. What was I supposed to do or say to my dad, who was sitting out there?

I did what many of us end up doing: I rolled up a large wad of toilet paper, stuck it into my underwear, and prayed no one would notice. I made it through the game, and as soon as we got back to the house, I showered and repeated that process with new clothes. I got no support from him and was ashamed of it. There was no room to be a girl with my father.

Many of the challenges I faced in dealing with my father impacted me from a mental and emotional standpoint. I had a lot of self-doubt when it came to my ability to perform well in school. There were even times I sought other male attention, even if it wasn't positive. Much of the treatment I experienced growing up with my dad bled into my social interactions and how I perceived myself. For most of my adolescence, I struggled with the thought that I was fat and ugly, and I shut down when I was scolded by other adults. I had many internal conflicts that I didn't know how to process mentally.

I had grown up working alongside my dad on his residential construction projects and learning what it meant to work hard. It was always difficult to navigate his abusive attitude and no-excuses mentality. As kids, we were used to tearing old carpets out of houses, repainting interiors, and even breaking up brush for some of his clients. Many summers saw us relying on water bottles for hydration and old McDonald's coffee cups for relieving ourselves. The work we did with him was brutal but useful down the road as adults. During many weekends and visitations, the three of us kids—always anxious and suffering from pits in our stomachs—tried to stay warm in the winter, avoid fleas in the summer, and not get sick from the plastic my dad used as walls to separate us from the dead pigeons on the other side of the house he was always "fixing up."

As we tried to make sense of the chaos and overall confusion we felt around our male parent, we sought escape many nights through our rental movies, my Walkman's cool techno beats, and radios that picked up random phone conversations off Raymond Street.

My bed was a stained futon from the garbage, my sister's bed was a wide array of pillows and couch cushions, and my brother slept with our dad in the only bed in the house. In the winter, my sister and I would lay a heated blanket on top of a pile of cushions in her room, then add many layers of blankets under and over us to retain heat. Summers were a whole different challenge, with the exposed floor framing and rancid carpet filled with fleas. My father's solution was to spread a baby powder substance all over the floor. Every weekend, we would return home to our mom covered in flea bites and rashes.

Even after several attempts, my mom couldn't get Child Protective Services (CPS) to investigate the home we were legally obligated to visit every weekend to see my father.

When we were home with my mom, she took out on us her exhaustion as a teacher and her frustration as a mother. Most of the time, it felt like she was too busy to help us, and much of the time, she would yell or be overcritical. There wasn't much in the way of physical affection or demonstrations of love. Emotionally, we all were hanging on by a thread.

Over time, though, my mom sought therapy so she could develop tools she could use to navigate her life and behaviors. My mom was our safe space, but it came with a mix of emotional and mental constraints. Much of the time, I, as the eldest, was in charge of taking care of my siblings. I made sure laundry was done, meals were made, and, at times, the house was tidy. I quickly learned to be an adult in many ways.

The plus from it all was that by the time high school came around, I had enough skills in hand to make some cash on the side to buy things. This was my first real start and feeling of financial freedom. With my mom's old red push lawn mower, I collected customers around the neighborhood and made some cash. I was thrilled to make some money and be able to shop at Hot Topic for clothes I liked (for the first time) that were brand new—not hand-me-downs from friends or family or inexpensive purchases from Dollar General.

FEELING LOST

By the time I started my freshman year of high school, the pit in my stomach had grown beyond a physical one. I was a nervous mess every time the weekend rolled around and visitation with Dad was upon us. I spent several weekends with him shortly after starting high school. Then I decided—after many long, tiring, and emotionally exhausting weekends—to stop seeing my father altogether. This was then, and to this day still is, the scariest thing I have ever done. I still remember shaking on my mom's front porch as I looked him dead in the eyes and told him I would no longer be visiting him.

This opened myriad other mental challenges and began the process of learning to accept myself. You can imagine how challenging and questioning a time high school became as I started to explore who I was and what it meant to be a girl becoming a woman. I liked boys, but I also liked girls. There was a mix of roughness and edginess to my personality, along with a desire to be soft and vulnerable. The increasing confusion led me down a path of fighting and hanging out with other outcast teenagers. Black became my dress code—a shield of sorts but also a badge of acceptance that said, "I am different." Even though I was one of the few freshmen placed on the varsity team, I gave up volleyball, which I had played reverently since fourth grade. I started to fight a lot more with my mom, fueled by my bubbling emotions and the unresolved confusion brought on by my upbringing.

Interestingly enough, skipping class became my salvation in a way, offering me an opportunity to stumble into a course redirect. In my second year of high school, I was a full-fledged bad kid with a reputation. My mom was worried and said that she felt like she couldn't control me or the decisions I made.

START

Jessi in high school at family holiday

I started to figure out how to leave the house in the middle of the night so I could go down to Madison Avenue to race or hang out with the street racers. I thought this was the coolest thing and the closest I could get to feeling like an adult. Sneaking out of the house was a way to control my life. It was a way to feel powerful and delve into the adrenaline that freedom created.

I especially remember one time I snuck out on a Friday night during the summer. My friend Destiny was having a sort of romantic fling with a guy in his twenties. He had a nice, big ol' white Ford 1500 with the quad back. She picked me up that evening in the back alley. I slowly opened the metal storm door and slipped out, doing everything in my power to make sure it didn't make a peep. Then I made my way outside.

We rode over to her guy's place, where we all hopped in his truck, then drove down to the west side, near the Indianapolis Motor Speedway, where there was an event happening. I was sitting in the back of his quad truck with the window open, wearing a little cream camisole and jeans, and feeling grown. Men would come up to the truck since the driver was moving at a slow crawl. They handed me beers and called me cute. The adrenaline was a rush I hadn't experienced before. These men either had no idea I was a high school kid or didn't care.

After a lap or two around the event, we headed back up toward Madison Avenue, where we met up with a bunch of street racers at Steak 'n Shake. People were already clearly intoxicated, and a wide variety of men and women were standing around oohing and ahhing at little Toyotas and old Mustangs. We grabbed some food and watched from a short distance as people lined up at one light and fully gassed it down to another light. The warm summer night air and the confidence boost I was feeling from being with these more mature adults gave me a feeling of elation—a sense of being okay.

This always lasted for a few hours, until I snuck my way back inside. This night was a closer call than usual because, by the time I got inside, I could hear my mom's boyfriend getting ready to go to his job as a chef. It was well past four a.m., so I dove for the couch, slid a few blankets over me, and faked it until he left the house a bit after five.

START

Instances like these gave me room to be curious and explore. I was so confused and unsure of what I was supposed to do in life. *What do I do next?* I wondered. *I am supposed to become an adult. What does that look like?* These were the questions running through my head every night as I yearned to be independent and **build a life that was different from what I saw with my mom.**

My dad had mentally abused us for years with his yelling and undermining treatment. At times, he was physically abusive, using a belt or a switch made from a tree. My own father stole money from me. My dedicated soup can (a.k.a. **piggy bank) would get filled from doing paid tasks one weekend** and disappear by the next. Nothing felt like it was my own or in my control.

Over time, these pieces of trauma showed up in my behavior and cascaded into my relationship with my mom. She did her best. But as a kid, I'd felt betrayed and unsafe. Sneaking out of the house and skipping class was a way to reshape the feeling of being unsafe and create what I thought safe was. In reality, I was **putting myself in harm's way as I naively attempted to find peace** and balance by controlling my independence.

At sixteen, I was a full rebel, dressed most days in head-to-toe black with ripped-up T-shirts, colorful fishnets, and a chain wallet. I hated school, didn't like my life, and continued to feel lost even with my mom's attempts to get me to therapy. Mentally, I was struggling to understand the societal norms expected of me and manage my unresolved past traumas.

Additionally, I was grappling with my sexuality. I perpetually found myself in a loop of asking myself, *Am I bisexual, a lesbian, or just confused?* One day, when my mom and I were in her little tan Honda Accord, I gathered up my nerve to share my confusion. Right in front of a red light, I blurted out, "Mom, I think I'm bisexual."

Looking a bit stunned, my mom calmly asked some questions. I unloaded previous experiences, including a huge crush I'd had on a classmate in middle school (Sarah, with her pretty brown hair and brilliant smile). Currently, I was being encouraged by one of my peers (Ashley with her freckles), who I was thinking of asking out.

I could see some of the fear etched on my mom's face. She started to go on about how much harder my life was going to be and said I needed to be careful. Today, as a grown adult, I better understand where she was coming from. But at the time, **I felt like she wanted me to fit in a pretty box and act "normal."** It was a hard conversation and a vulnerable moment in my **growth.** A little relieved to have shared, I tried to offer my mom insights about my current crush. We talked about Ashley for a bit, with my mom sounding a little skeptical about the quality or longevity of that relationship.

Eventually, convinced my mom was out to get me, and feeling stuck, trapped, and unsure of all the emotions bottled inside me, I did the most foolish thing of all.

I ran away from home.

I packed my favorite black Hot Topic bag with essentials, put on my black trench coat, and left my mom a note. I had already made plans to stay with a girlfriend who had older siblings who were more lax. My plan was simple: hang out with her for a while, find a job, and get a place of my own. Working hard was never a problem for me. The problem was finding where I fit and figuring out what control I could have over my future.

I successfully made it four whole hours before my mom tracked me down. She had gone around to places she had dropped me off at in the past, including an ex-boyfriend's place. He was the one who knew who I was staying with.

Admittedly, the whole thing had given me a huge pit in my stomach, but I didn't know what else to do anymore. I was fighting with my mom every second we saw each other; school was a disaster with kids bullying me, and I was getting into fights. I yearned so much to be an adult, make my *own* decisions, and feel in control of *my* future. But my running-away stunt scared the living crap out of my mom, and I felt the wrath when I got home. I lost every privilege and had a regimented schedule that included my grandpa picking me up at school. There was no more Jessi roaming free. Instead, I was in full lockdown mode.

Still pissed off and rambunctious, even with my mom's rigid management, I kept acting out at school. I found ways to hide out in the stairwells during class, get home, and delete the message from school saying I'd missed whatever period of science. I even found a small gaggle of kids to skip with.

One day, I was hanging out with a few of these anxious teens when we ended up in a wing of the school that was being worked on. There was a bunch of drywall placed as a protective barrier between where we were and the room next door. And just like a smart-ass, I asked Robert, my mohawk-haired and pierced buddy, to hand me a Sharpie. On one wall, I drew my signature heart with wings and a devil's tail. I thought it was the most hilarious thing. Of course, I didn't consider that there were cameras around or some way I could be caught.

Two days later, I was called down to the main office. Sitting there was Robert, with an "oh shit" expression. I sat next to him and asked, "What's up?"

He shrugged and said, "Don't know."

A few minutes later, I was pulled into the vice principal's office and immediately peppered with questions about some gang. I was genuinely confused. She asked what gang we belonged to, and I shook my head, telling her that neither Robert nor I belonged to a gang. Eventually, she got to the main point, which was the graffiti I had drawn on the drywall. She flipped a piece of paper over to reveal a half-erased, washed-out version of what I had drawn on the wall a few days prior.

She was pissed at both of us, so I redirected her anger at myself, pointing out that Robert had nothing to do with it. He'd handed me the weapon of choice (a.k.a. the black Sharpie I'd used to draw the winged heart), not knowing what I was planning to do with it.

START

Robert was let go, and I was immediately escorted to a bathroom and handed a pee cup. (FYI: *This is a big no-no without parental consent.*)

A little dumbstruck, I went in and did my business, unsure why this had anything to do with the graffiti and confused about what the big deal was. The drywall was temporary support.

Eventually, my mom showed up, and I was placed into a full week of in-school suspension. However, this suspension was unlike the typical punishment. I was tasked to work with the janitorial and maintenance crew for a full week. While my peers went to class, I was going to be cleaning graffiti, replacing broken railings, and patching holes in walls. Those skills from working with my dad were handy indeed.

Day one was a bit nerve-wracking. I had no clue what to expect. It quickly became a full day of walking around and being chaperoned while cleaning, painting, and repairing.

Honestly, I loved it. I liked doing stuff with my hands, and the days went by much quicker than they did while I was in class. After a full week of working with the head custodian, I had a new perspective on what they did and the impact of my actions, as well as those of my peers.

Things took a new direction not too long after when I stumbled into my friend's welding class while skipping yet another boring class.

CHAPTER 2

WELDER

"Sir, excuse me."
"Hey there, man."
"Oh, you don't look like a welder."
"I would never have guessed you were a welder."

These were phrases I constantly heard when I was stepping into a gas station for an iced tea or on my way to an event with new people.

What does a welder look like? Many assumptions based on how I dressed, appropriately for my trade, made others question my sex. Because I picked a certain trade for a career, I no longer fit in anyone's box.

Regardless of my rough upbringing, I still reflect heavily on the skills I was exposed to. Even now, I consider how unique this makes me today. I have a set of tools I can utilize for better interpersonal development and relationship building.

At sixteen, I stumbled across welding one day while I was skipping class. I mentioned skipping to my friend Becky, and she invited me to hang out with her in the welding class. Feeling fearless, I said sure.

When we went in, the first thing I saw were these concrete booths. She hurried me into one and pulled the plastic curtain closed. When she came back, she handed me a bulky gray thing to put over my face. It had a headband and a face shield that flipped down. Becky called it a welding hood. She got set up and told me to flip the hood down. Once she struck the spark on the metal, the flames of my first love—welding—began.

As Becky struck the first arc (stick welding) on the horizontal plate, I watched in amazement. This was something I had never witnessed, and I quickly became fascinated.

Why has no one told me about a trade option? I wondered. *Also, why is there only one girl in a class full of boys?* These questions started my drive to learn and do more. Shortly after, I figured out a way to take night classes through my high school. This became quite a responsibility.

After feeling aimless all of middle school and now high school, I'd finally found purpose. This shifted my attitude toward school and reframed my need to do well and graduate. The night classes kept me out of trouble and gave me access to one of the best mentors, Dan Johnson (may he rest in peace). For the first time in my life, I felt like I had a male role model who treated me as an appreciated human. He was patient and kind and helped me refine my passion.

Infamous black sweatshirt

The night courses were dual credit, which meant after graduation, I would already have college credits through the local technical school, Ivy Tech Community. This opened my eyes to many new possibilities and started to build my confidence. Welding became an outlet for me to fuse my love of art (sculpture work) and still have a means of taking care of myself as an adult. #Independence #FinancialFreedom.

In my junior year of high school, I was named the Rolls Royce Welder of the Year. This gave me continued fuel in my journey.

SHEET METAL SHOP

At the end of my sophomore year, with the help of my welding instructor, I landed a summer job as a welding helper at the local sheet metal shop, Steelcraft Metal. This was a union shop that paid me chump change to do odds-and-ends welding jobs, assemble HVAC framing, and push a broom. I was excited to get this opportunity, which felt on par with my male peers. In a way, it was a form of acceptance—like an equal almost.

What I didn't expect were the horrible conditions or the unacceptable behavior. What can I say? I was a teenager with a limited lens of the world.

On my first day, I drove up in my used pearl Nissan Maxima, my pride and joy gifted to me by my grandpa, who sold cars as a side hustle. I stepped out of the car, very nervous and unsure where to park. Wearing my new black Dickie pants, black T-shirt, and skull bandana, along with my welding gear, I walked in the front door. I received no warm welcome. Instead, I was shoved into an office to fill out some forms.

The manager of the sheet metal shop was an angry-looking dude who, even at seven a.m., was already on the phone yelling at some poor soul on the other end. I was standing in a dark office that looked out at the main work floor, with its large windows, at the front of the medium-size building. Below was a large, open, dingy space that looked like a garage. There I saw welders, grinders, and a few metal cutting stations. The smell that sifted through the office door was a mix of dirt, oil, and rust. Not knowing exactly what I was going to be doing, or even what this place did, I hurried through my paperwork. Once finished, I was escorted to the main floor below.

Everything around me was buzzing with noises and the smell of sweat mixed with tobacco. I was quickly shown where the bathroom was and then walked over to a large metal table. Standing there, slightly dumbfounded because my tour guide was not communicating anything, I nervously waited for some semblance of instructions. The man walked away and came back with this long, straight piece of HVAC setup and leaned it against the table. He mumbled something and then asked if I knew how to MIG[1] weld a straight line.

I responded yes, partially quizzically.

"Great. See this here?" he asked, pointing at a skinny gap from the top of a long rectangle where the metal overlapped ever so slightly. "Weld this seam shut."

That was it, the only bit of direction I was offered.

Still feeling very unsure about what was expected, I set my stuff up and looked around me. Men of varying ages were all busy getting their areas set up for the day. I put my gear on, set my welder, and went to work. Oh so carefully and somewhat shakily, I welded up the seam on the HVAC section in about thirty minutes. Even though it felt like an eternity, it was a

---

[1] MIG, Metal Inert Gas *Mig Welding: The basics for mild steel*. Miller Electric. (2023, May 26). https://www.millerwelds.com/resources/article-library/mig-welding-the-basics-for-mild-steel

pretty simple task. No one was really paying attention to me or working in my area.

Unsure of what to do next, I found a broom and started to clean up around my workspace, hopeful that Mumbles would come back and give me more work or direction.

A little before lunch, the escort from earlier showed up and asked if I was done. I said yes and asked him if he wanted to take a look. He mumbled something again and took the section down to another arm of the shop.

A few moments later, he turned and said, "We don't have a ton of work right now, so I will introduce you to a few folks." He went on to let me know that my trainer would be in the shop the next day.

After some quasi-intros, I was shepherded back to my table, where a few smaller HVAC sections were lying on my worktable. He told me to work on them for the rest of the day.

I was flabbergasted at how little there was for me to do. Unsure whether they didn't have faith in me or were a failing shop, I kept quiet and did my best to look busy. This included cleaning out my work bag and organizing under the large metal table.

One of the things that quickly became apparent was that everyone in the sheet metal shop smoked or chewed tobacco. The second was the number of posters and calendars that displayed partially nude women.

Taking a few glances around when I traveled to the bathroom, I started to piece together that the main focus of this place was to build HVAC ducts and metal framing. Simply put, the shop would start with a large flat span of thin gauge sheet metal, (such as stainless or aluminum), then bend it to form square or rectangular ducts. Those sections would then be sent out to be welded by various stages and then inspected.

I managed to slowly work through my task at hand and be "busy" until the end-of-shift whistle. I was thrilled to get out of there but not so excited about returning in the morning.

The next day came quickly, and it took all my courage and might to return. There was nothing warm or gratifying about the environment I was in, but I wanted that boost of experience so badly that I swallowed all the gut feelings and red flags.

As soon as I entered the shop, I saw an older-looking man at my metal table. He introduced himself as my trainer and gave me a side-glance. Looking at his long beard and scraggly hair, I was a little intimidated. But did my best to present myself confidently. He spent time walking me through the daily activities and expectations. He even explained the layout of the shop. After an hour or so, he asked me some questions about my skills and what I was comfortable with. Soon, he was placing different ductwork pieces on the large metal table for me to weld.

One thing was for sure: this sheet metal shop was struggling. There were men just sitting around talking to one another. The space was quiet and felt empty. Hardly any motion throughout the shop.

Over the next few weeks, my welding trainer took a shine to me. He would joke with me and even offer me a cigarette from time to time. It felt like a badge of honor even though I was a few years away from the legal age.

Even when we were slow, with no welding work, he would keep me busy. At one point, he had me learn how to run quality assurance tests on HVAC seams. This was fun. I took a special dye marker and ran it on the outside of the welds. Then, I crawled inside the units with a black light and marked up spots that had burned through or weren't welded all the way.

The shop continued to be slow, and it was getting harder for me to appear busy on the weld side, even to the point where when I was doing TIG[2] repairs, my trainer told me to move slower, which I didn't understand. Eventually, I was moved over to the right side of the shop. Visually, the shop was pared down the middle and split into four sections. Framing was far right and farthest in the back, by the delivery doors.

I was put on framing duty. That meant cutting toxic metal frames made of galvanized steel,[3] which leaves white whispies floating in the air when welding or cutting is happening. The little airy whisps are zinc. With prolonged exposure, they are incredibly harmful.

---

2   Tungsten Inert Gas. Pfaller, A. (2021, September 1). *How a TIG welder works and when to Tig Weld*. Miller Electric. https://www.millerwelds.com/resources/article-library/tig-it-how-a-tig-welder-works-and-when-to-tig-weld#:~:text=TIG%20stands%20for%20tungsten%20inert,the%20tungsten%20and%20weld%20puddle.

3   Galvanized steel and zinc poisoning. *Avoid galvanize poisoning when welding*. Red Steel Manufacturing. (2023, May 2). https://www.redsteelmh.com/avoid-galvanize-poisoning-when-welding/

At the time of my summer to-work program, I was taking night school classes to blast myself through senior year so I could graduate sooner and start making real money faster. By the end of my first month, I felt ill all the time. I was headachy and nauseated, and my sense of smell was off. I had a hard time concentrating, and I was seriously fatigued. At the time, I had no clue about the hazards of the framing work and its impact on my physical health. The zinc (the white whispies) was poisoning me. Without proper ventilation or a respirator, the exposure would continue. Most of the time, I played it off, assuming I was just tired from my early morning work schedule in combination with my night school classes four evenings a week.

This was a testament to how stubborn I was, regardless of how terrible or difficult the situation was. I was committed to finishing my summer and completing my night school classes.

A few weeks before the end of my twelve weeks of working at Steelcraft Metal, the shop was dead, and I was moved over to cleaning duty. This meant cleaning disgusting men's bathrooms and learning how to vacuum out the presses. I was stuck with a man who was prickly and didn't want me on his turf. He managed three separate robotic-type presses and lathes, which formed many of the HVAC pieces and parts that got sent out to be welded. Since there weren't any orders, he was cleaning all the machines. He showed me how to take vents off and vacuum, gave me a broom, and told me where to sweep. It was a semi-sad way to wrap up my last few weeks, especially when I was hoping to weld more than I did.

When my last day came, I was intent on leaving a few messages for two main men I worked with. One was my welding trainer, who happened to not be there. I welded him a thank-you message on some scrap metal. The other was for the framing manager, who was just mean to everyone. So I welded him a message that said, "You are a dick," and left it front and center at his workstation. I then had the framing crew grant me one wish before I left—being wrapped up in the heavy-duty Saran Wrap-type plastic wrap we used to cover the HVAC sections when getting them ready to be shipped. My last day was my final middle finger to the bad experiences but also a rebellious way to rejoice at the end of my punishing work experience for the summer.

I completed my time, and my sad paychecks reflected the last few weeks of what was left of my summer. Working at Steelcraft Metal was an eye-opener to the world I was stepping into and how rough it was going to be. However, being the stubborn (young) woman I was (and still am) meant I continued onward regardless of the push to make me quit.

GRADUATION

I'd rapidly burned through my junior year of high school. And because of my extra dedication over my summer, I was almost done with all my requirements to graduate by the time I got to my senior year. I just needed to complete two final credits. For fun, I took an art class, which I attended two mornings a week. The rest of the time, I worked as a cashier at Walmart

Market. I was restless and ready to be free to seek a new level of independence and carve out my path as a welder.

Upon graduation in 2007, I immediately turned around and enrolled in Ivy Tech Community College, anxious to earn my degree while I continued to look for a full-time welding gig. I continued working for Walmart and even got promoted to customer service manager to help pay for classes and my first apartment. I was on the verge of eighteen and bursting to be free.

After bouncing around a few odd jobs with Walmart and Target while I worked toward completing my welding certificates at Ivy Tech, I finally took a position at Hawthorne Auto Works Body Shop in Fishers, Indiana. This was my first big-girl role where I could take all my tech training (auto and welding) and crawl around a truck bay, replacing bumpers, spot-welding bodywork,[4] and learning the operations. I decided to focus solely on the work at the body shop and put my technical courses on hold after the shop manager promised to enroll me in the certification program.

Over time, this position became an experience similar to my Steelcraft Metal role. I was the only woman in the shop. And because of how inexperienced I was with bodywork, my training

---

4   Bodywork is the repair of the overall appearance of a car using different methods, such as fabrication to reshape metal and Bondo to fill in minor dents. Or replacing entire panels and other necessary repairs all the way up to the paint stage.

THE TRUTH ABOUT BEING A WOMAN IN CONSTRUCTION

was slow. Eventually, I proved myself worthy of being trained under the lead body tech for Geico. I had also started working on hot rods and classic cars on the weekends with a buddy. This was my attempt to learn more and find my niche.

I felt so powerful and badass with my ability to pull my tools out and work on my truck without guidance. Even walking into AutoZone and picking up parts and supplies felt like an achievement. I knew what I needed and how to install or upgrade it.

Jessi with a Hotrod

Both experiences—at Hawthorne Auto Works and with my buddy—were informative in several ways. I had some kind teachers who were encouraging and accepting of my interest.

WELDER

When I finished up my required detailing tasks, I'd walk right over and offer to help as one of the older men in the shop was tearing down a car or replacing a window. Fred was especially kind and let me borrow his tools as I took off a part of a door panel or replaced a mirror on a vehicle. Sometimes, he would even let me watch off to the side as he mixed up Bondo and performed bodywork on a hammered-out piece of steel that once was a wheel well. He had an old-grandpa vibe and moved at his own pace, which never felt rushed.

On the weekends, I would practice my newly learned skills on my own green 1994 Dodge Ram 1500. Every so often, I'd be able to pull from the body shop less damaged replacement parts for my truck, including a new hood that wasn't rusted out.

Over time, I replaced the battery, started to repaint the interior black, and upgraded the brakes with more robust performance-quality parts. I rode high, feeling proud of the slow improvements.

Interior of my 94 Dodge Ram/ Jessi working on bumper

During the summer and early fall, it was always fun to travel to the different car shows. I would walk around in a sea of people, oohing and ahhing at the various vehicles. To me, they were captured works of art, with blood and sweat poured in by car enthusiasts. Many times, I would walk up and be able to identify a car or truck by its year, make, and model before reading the sign. There were moments when I would take a car owner by surprise—most of the time a man—by asking for specifics about his 1968 Stingray and whether it was the original color. Of course, as proud owners, they would be excited to share more details and talk about the journey it took to restore their treasured investment. These sweet experiences helped me discover my strength and ability.

I felt power in the knowledge I was gaining.

As much as I was enjoying all that I was learning, significant barriers started to surface. The spaces I was able to learn in soon became hardships. Mentally, I was struggling with body issues as I received ongoing comments from my "buddy" regarding my physique. He would over-compliment my features, such as my butt, or say, "Look how small you are. You fit perfectly in the trunk." Sometimes, he compared my skills to those of men. It was toxic, but I was so desperate to learn that I pushed past it as much as I could by ignoring it.

Hawthorne Auto Works became a closed door. My hopes of becoming a certified body technician started to seem more like a pipe dream. The weeks turned into months without any certification courses being offered. When I asked, the shop manager, who'd made the initial promise to enroll me, gave me constant excuses. She'd say, "No, not yet," and, "The timing isn't right." My official trainer ended up having zero patience to train me. He even went behind my back to the shop manager to report that I was planning to learn as many skills as possible and eventually open my very own shop.

This claim came back on me, and they punished me by putting me back on the cleaning crew. I was stunned. All the effort I had put into working with the different teams, from the painters to the front desk, and even volunteering with other body techs meant nothing to my trainer or the shop manager. Instead, they put me back on "woman's work," cleaning the bays, washing the finished vehicles, and vacuuming the front of the shop, where the customers came in and waited.

After months of going back and forth with Hawthorne Auto Works and their false promises to train me seriously as a body technician, I decided to start up my Ivy Tech courses again. Feeling disappointed after working so hard and not getting what I'd wanted, I took a role at Tucker Body Shop in Carmel. They offered a little bit higher pay and a large promise to train me. This experience ended up being a repeat of being given empty promises and used as cheap labor.

SPARK OF LIGHT

After much difficulty and reflection, I took a big step back. I ended up bouncing around a bit, working for AutoZone and eventually finding my first full-time welding gig at ILLI in

Westfield. By this point, I had graduated from Ivy Tech with the equivalent of an associate degree focused on technical welding. This was an eye-opener to a whole new industry and the manufacturing world. Soon, I was being trained on MIG robots and learning how to pull jigs and weld various automotive parts and framing.

Jessi Welding

My role initially with ILLI was focused on bus frame accessories, tube framing, seat pedestals, and seat belt constraint brackets. Most of the steps were fairly simple, involving pre-cut steel materials labeled in a bin that was placed inside a frame contraption called a jig.[5] The jig was a series of clamps and outlines of the metal parts that was designed for easy assembly. The hard part was having the welding skill to fuse the parts in the correct locations using a MIG, or Metal Inert Gas, more technically known as Gas Metal Arc Welding (GMAW). Simply put, a MIG is a tool that looks like a gun and feeds a metal wire and electric current to create the welding itself (a.k.a. fusion of metal).

Over time, there were complaints about the air quality of the workspace, as the fumes that circled all of us as we worked were incredibly unhealthy.[6] A few of us, including me, were selected to wear air monitors to track the contents and particulate matter each day. I even wore a medical mask under my welding hood to create a barrier that turned black by the end of each shift. Since we saw no changes in the workspace after the monitoring and the welders and operators weren't told one way or another, we assumed there was nothing harmful or the levels were considered acceptable.

Time passed, and my work and skills developed—so much so that the other two women I worked with started to keep a

---

5   Metal contraption that lays out the pieces and parts in such a way as to take the guesswork out of assembly. Usually very heavy with clamps.

6   Unhealthy air quality typically occurs from the lack of ventilation, causing the individual to inhale harmful metal fumes and gas by-products - Controlling hazardous fumes and gases during welding. (n.d.).https://www.osha.gov/sites/default/files/publications/OSHA_FS-3647_Welding.pdf

distance from me and even talked behind my back. I focused on why I was there, which was to make a living, learn, and grow my skills.

I kept my work gear and welding backpack under the table I worked at. This had been my storage for months without any problems with fire. After all, my welding bag was made to handle sparks.

It was not made to handle a *fire*.

Over time, things escalated with the women to the point where one of them put a giant box of Kleenex next to my work gear and backpack under the table. The box caught on fire and burned all my gear while the two women did nothing more than stand there with semi-panicked looks on their faces as I put the fire out.

My time in the back of the welding shop had run its course, and I felt like it was time to advance. After months of what I considered the "easier," less challenging welding projects, I asked to move up the line to where the men worked.

Yes, we were segregated. Women worked on one side of the shop, doing light-duty welding, and men were on the other side, doing the heavier tasks. After having a few conversations with the lead shop manager and discussing my interest in doing something new, I was moved up and started running frames.

The environment itself had already taken a toll on my health. My eyesight was starting to become spotty, and breathing was a challenge. The space we worked in was large but heavy with fumes from the robots, various machinery, and welding operations. The vents were well over twenty feet up and nowhere near the welding areas.

I did my best to keep moving fast and flip frames on the line. We even had a running tally that showed how many frames we completed each shift. I quickly adapted and learned every part of the line. After months of struggling to get comfortable and find a way to fit into this company and team of welders—in addition to developing tendinitis and carpal tunnel syndrome—there seemed to be no light in sight at the end of the tunnel.

My health started to decline, making it hard for me to physically handle the repeated work and overuse of my body. I ended up being let go from ILLI after many poor attempts to accommodate my abused body. This turned into a lawsuit that took several years to settle and started me down the rabbit hole of self-doubt. I began to question my career path and whether I was cut out for the work. Not once did I consider that the environment itself might have been the problem and been intentionally not inclusive.

REFRESH AND PIVOT

A year later, I found a welding gig with NexFab Manufacturing. This was an ideal role with more challenging duties, including TIG (Tungsten Inert Gas) or GTAW (Gas Tungsten Arc Welding) welding stainless steel and aluminum. This process is similar to MIG, except that you have a pedal that helps drive the current through your torch. And instead of a trigger that metals a rod, you have a piece of tungsten that creates the arc or metal puddle that you feed, by hand, with a long, skinny metal rod. The company did a wide range of work, including small- to large-bore piping, medical equipment, and phone boxes. I was excited to do something new and work with the smaller team I would be on. It was still shift work, but the environment was cleaner and had individual vents set at each station. Thrilled to be taking on cleaner and more challenging work, I dove in.

 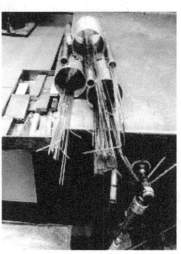

Jessi with hood and matching nails *Jessi's custom rod holder*

Over time, I was given many new responsibilities, such as training and hiring new welders. There were a few women on my shift, including one welder and a machine operator. The other female welder came from foundry work and was struggling to gain skills quickly. She was incredibly small and did her best to be even smaller in the presence of others. I did my best to be encouraging and help her with the challenges she was facing.

She started to seem a bit more confident but still stayed in the confines of what she was told to do—nothing more, nothing less.

My pal out on the floor was the opposite. She was a mom of three, and I nicknamed her Barbie. (I even gave her a rubber bracelet as a birthday gift that had the word *Barbie* in pink.) Outgoing and sometimes very loud, she always had her nails done and was there to make money. She had one of the dirtier jobs in the machine shop: flipping sheet metal under the laser cutter, then using a hand sander to knock off tags. Sometimes, she would even need a skinny triangular hammer tool to get the cut pieces unstuck. Barbie didn't care what others thought of her, and she dressed in jeans and bright colors every day. The only woman on the first shift, she didn't take a lot of crap from the others on the floor. In return, they left her in peace.

I finally felt like I was part of a team where I was earning respect, making more money, and trying out new skills. I learned how to operate some of the metal presses and plastic welding,[7] which is a fancy way of saying I monitored a machine that uses molds to make plastic parts. Sheet metal was a big function of our work, so it was important to be strong enough to lift sheets on a shear and utilize buffers or grinders to take off excess or spotty tags that were sharp.

---

[7] Plastic welding is the process of melting plastic into a mold or joining two piece of plastic by melting the surface. Usually uses a mix of pressure and heat.

I worked closely with our quality assurance and quality control team to drive efficient processes and offer new methods to help reduce human error. Mainly by staging plates and the hardware a certain way, I was able to help the floor team move more efficiently with fewer issues. The panels that were typically prioritized were stainless and had metal studs that had to be pressed in with a special spot-welder. By tearing off only a specific piece of plastic covering and rotating the panel a certain way, we made the process seamless, resulting in almost no defects. Interestingly enough, the lead welder ended up getting credit for the change. He received a plaque and even had his photo put in the hall. At the time, I wasn't overly worried about it. The work I was doing was all about improving the team as a whole.

The day shift was a dream. That changed when I went to the second shift. I had a new supervisor who micromanaged and had no idea how to take care of his team. Additionally, I was hearing stories from other women around me about the behaviors of our shop owner and how he was physically inappropriate.

I started to document the challenges I was seeing, including the verbal abuse I was facing with my new supervisor. He was always running around with his head cut off. His frustration seemed to bleed out onto me—one of only two women on the shift and the only welder. Then, other bad behavior surfaced from other welders as well. Tools would go missing, stations would be left a mess, and even basic things—like communicating what task was left unfinished—became difficult. My job became incredibly hard to manage, and I again felt alone in trying my best to figure it out.

Eventually, I called it quits after a little over a year with the company. In my exit interview, I outlined all my challenges—including seeing a male welder be recognized and *awarded* for an efficiency process I had created and trained others on. To top it off, the company didn't seem to be doing too well, to the point where the welders were getting pulled to paint, clean, or work in assembling departments.

Unsure and unclear about what to do next, I started to explore other options. *What do I do now?* I wondered. *Was this all a mistake?*

Maybe I wasn't cut out to be a welder after all.

CHAPTER 3

CONSTRUCTION MANAGEMENT

Figuring out how smart you are can be an awakening. I frequently heard that I was "too dumb." Sometimes, someone would say to me, "That's a hard job. You sure that's the path you want to take?" But what I heard was, "I dare you." It was a sort of mantra I have always rebelled against as I broke barriers.

At twenty-four, I felt like my life was in shambles. After a series of bad experiences working as a welder and a relationship that came to a crashing halt, I felt like I had hit rock bottom. After some serious deliberation with my mom and sister, I packed all my belongings, cashed out the tiny bit of retirement to my name, and moved to Muncie, Indiana. This was where my life redirected itself. The pieces that shattered started to come together into a new picture.

My mom quickly came to my side to support me. Together, she and I agreed to find an apartment and start a path to my undergrad studies at Ball State University. She was living in a tiny studio at the time and feeling isolated. After some discussion, we both agreed this was a good opportunity for us to support one another. Within days, we found a reasonably priced three-bedroom apartment close to campus. We loaded up both of our residences and got out of town as fast as possible.

My sister came down from Purdue University, where she was working at a lab. Together, we all tirelessly packed and then unpacked my mom's and my belongings into a new place for a fresh start.

At the time, I was racking up the transferable credits at Ivy Tech Community College which I knew had a partnership with Ball State University. The plan was to use those credits to knock out a bachelor's degree in a matter of a few years.

The original plan was to pursue architecture, and I was excited to plug into a more intentional creative path. But I quickly realized that many of the concepts were focused on ideas and weren't necessarily based on hard, technical knowledge. At the end of my first semester at Ball State, I redirected my major to construction management and chose sustainability for my minor.

This was the eye-opening experience I had yearned for all these years as a welder. It was the next level to my comprehension

CONSTRUCTION MANAGEMENT

of the trades and skills I'd picked up doing residential construction in my youth. Classes made sense. I was able to dive into the technical details while providing new insights based on personal experiences from welding. The environment itself was familiar—a repeat of a male-dominated space and a myriad of opportunities to explore.

At first, I felt accepted into the program and got lots of support from one of my professors, Mr. Mezo favorably nicknamed M2. There were also a few women who advised the construction management program.

Jessi with M2

Jessi being awarded for top achievements

I attended the women's lunch and learns, where we all sat together and discussed experiences or asked specific questions relating to the field we were going into. As a seasoned tradesperson,

I had more perspective than the average undergrad. I enjoyed my time sitting in a room full of curious and ambitious women. Together, we shared stories, asked questions, and built a sense of community as minorities in the male-dominated program. We had a space that was ours, where we could openly talk about subjects not well understood in mixed company. It felt good to sit around a table eating free food and sharing our ideas for the future while pondering what life after graduation was going to look like. It was one of the few spaces where we could connect and empathize in the warmth of each other's mutual understanding.

However, toward the end of the first year of my curriculum, the university imposed several new changes, and the welcoming experience I'd originally enjoyed became an isolated affair. Due to a program technicality, I was no longer considered part of the traditional construction management umbrella. Instead, I was placed under the Applied Technology department. I was taking all the same classes as my construction management peers, but my minor was sustainability instead of business.

Ignoring the new barrier, I still pursued other ways to practice my skills, such as participating in the Associated Schools of Construction (ASC) Competition. I joined a small team of four young men and one other woman to work through a grueling twenty-four-hour competition in Aurora, Illinois. Each team was sponsored by one of the universities that participated in the ASC event. A ragtag mix of us made up different divisions (R.I.P. Darian). My team was fortunate to be in the concrete division, which meant a more specialized focus.

CONSTRUCTION MANAGEMENT

Before we traveled up to Illinois, we as a team—all six of us, with our wide variety of experiences—had met a handful of times to prepare. A few of the team members had competed before but on different teams and divisions. Although we ranged in age and construction exposure, the one thing we had in common was our competitive spirit and drive to successfully apply our skills.

We were there to win. And we did all we could to create templates and figure out a way to stand out or brand ourselves.

On the first day of the competition, we assigned one of our hotel rooms as the field office. Everyone had a title. Mine was the project coordinator. We were given minimal information to pull together a proposal for a concrete project. Throughout the day, the judges occasionally stopped in to observe and answer questions we might have about the project.

By three a.m., our team had built out a deck to present and had proposal documents ready to print, then bind. Throughout the late night and early morning, we coordinated our summaries to tell a clear story of the scope that lent itself to a smooth sales pitch. The next day, we suited up and waited our turn to present.

Nervous and excited, we stood around and waited for our team to be called. When they called our group name, we took our places and set up our props. The fastest twenty minutes flew by, and we made our concluding statement. The rest of the day, we took the opportunity to nap and pack. Finally, later in the afternoon, we met up with all the other teams and the various divisions.

One by one, we watched groups get awarded first, second, and third place for their work. After some time, they opened the Concrete Division, and the judges gave brief intros for each team and what the scope included. Working from third place to first, they started the countdown.

Eagerly, we sat and waited. Second place was announced, leaving us in the final moment of wondering, *Did we get this?* To our surprise and glee, we were called up as the first-place winners. It was such a feeling of accomplishment. All that hard work as a welder and even working with my dad seemed to pay off in a moment. Proudly, we stood and accepted our award. The event was an encouraging experience that made me feel closer to my peers and like I had finally found my path.

ASC BSU Team Concrete First Place

CONSTRUCTION MANAGEMENT

This was it. I was meant to be in construction.

Riding on the success I felt with the Associated School of Construction, I ran for the Construction Management Student Organization President (CMSO) role. I'd joined as a member the prior year and had enjoyed it. It was another avenue to learn from others and boost my resume a bit. There were only a handful of us interested in an officer position. It came down to me and a young dude in my program. Competing, each of us claimed to be able to offer the best events and opportunities for our peers.

After a series of brief campaign speeches given at the end of a day of classes, members voted using little pieces of paper that were collected in someone's hat. The room was a mix of maybe ten students sitting at desks. There was a whiteboard with scribbled campaign notes for each officer role. Our current president finished counting and sorting the votes, then stood up and announced the newly elected officers.

My nervous insecurity caused me to sit and tap my foot. At last, I heard my name called as the next year's CMSO president.

Of course, already having some thoughts brewing, I started to line up biweekly CMSO events and work closely with the other officers. We had hardly any budget, so we learned to be creative. One of the events was a matching game built using only a basic Word document. The four pages had a series of stock photos and terms mixed with typical construction phrases. Teams of two tried to race through each section. It was quite the ordeal, watching the various reactions and excitement as individuals got confused or figured out the answer. At the end of it, we declared a winning team and awarded them construction swag from a local contractor and a gift card to the on-campus bookstore.

Celebratory women in construction event for undergrads

My term as president felt successful. We had tons of engagement at the events and developed innovative ideas on how to support each other. Some members were more focused on having fun, like doing a cookout on the quad, or offering midterm and final study halls to help each other tough out the anxiety of exams. It was a delightful way to explore my skills and learn more from my classmates in a new way.

INTERNSHIP

As I neared the last year of my undergrad career, I knew how important it was to acquire some resume experience that I could call related to construction management. So, in the spring of 2016, I applied to various companies and interviewed like crazy, trying to land the ideal internship that would allow me space to learn and wear many hats. It came down to two companies: Hunt Construction, known today as one of AECOM's subsidiaries, and BSA LifeStructures, an architecture and interior design firm.

On paper, Hunt appeared to be the ideal option. It was a large, well-known general contractor firm with various projects I could be exposed to. Initially, I had a few concerns after I interviewed. Each interview was with a man, and an older one at that. I was told that if I took the role, I would be in the office all summer working on their Autodesk Revit software focused on clash detection. This was not what I would call inspiring or exactly the hopeful job experience I was seeking. But Hunt did present itself as a soft landing with a decent internship wage. The conversations and lunch I had with the older gentlemen were encouraging but not in line with what I wanted to get out of my summer.

BSA LifeStructures was a completely different option. It was a design firm that had several in-house functions, making it very diverse in the work it managed. When I interviewed, it was with two men and two women. I was thrilled to be in a space where I could look around and derive solace from the female energy. This is why representation matters—you need to be able to see yourself in other roles but also have a sense of familiarity. The role was not a typical construction role. Instead, it was contract focused, and my role was offered as a construction admin intern. At first, I was hesitant because the pay was a bit less than Hunt Construction, but my drive was going to be shorter.

After some consideration and thinking about the feeling of "fitting" with BSA, I took the role. To date, this was still one of my best experiences. My summer with the design firm changed my perspective and boosted my confidence.

Jessi as a proud BSA intern

On day one, I was ushered into a nook library with a group of other interns. We were all walked through our onboarding, which helped us understand what the firm did and who we should turn to if we had concerns. Day two, I met the team I was working with. To my delight, it consisted of three women and one man. I was beyond thrilled. We all hit it off marvelously, and I was taken to lunch. I had a bar desk with a glass nameplate. I was flying high that first week and excited for the new adventure.

Over the next few weeks, introductions were made, and I learned more about the different divisions. I came to better understand AIA format and documentation, as well as the way drawings were referenced.

One of the contract admins invited me to go on a few site visits with some of the bigger key clients at their projects—one near Shelbyville and the other in Illinois. It was great to go out and see how she coordinated project meetings and talked to the folks building the work. I even expanded my network of other general contractors. After a few weeks of these interactions, I realized I wanted to explore more.

At the time, Dan was our contract project manager. He oversaw the work the contract admins did and coordinated details and client engagements. He and I had a great conversation one day, during which I learned more about what he did. He was kind and encouraging with my questions. Plus, he was patient with the novice level of comprehension I had at the time.

One day, I bravely asked if I could join him out on one of his site projects. He was always traveling around, spending more time out in the field and less in the office. It felt like a perfect

THE TRUTH ABOUT BEING A WOMAN IN CONSTRUCTION

opportunity to expand my knowledge and personal comfort zone.

At first, we did maybe one project a week. Then, after going on a few site visits and seeing my growing desire to learn, Dan invited me more frequently. Eventually, I was spending more time at site meetings than in the office. And that felt good. I was going all over Indiana, sometimes as far down as Bloomington to Indiana University and as far up as Notre Dame, to see these large projects firsthand.

Dan acted as my mentor, offering career advice, postgrad suggestions, and even detailed answers to my questions. He helped me see projects and construction in a new light, forcing me to see details otherwise ignored and the importance of staying on schedule. (Liquidated damages[8] are no joke!) I was learning so much about how drawings were noted and punch list items were tracked. And I was encouraged to ask questions about building details. I was buzzing with the real-world application and enjoying how my prior experience as a welder and in residential construction as a kid helped me communicate effectively.

By the end of the summer, I had traveled around three states, attended more than a few dozen project meetings, and walked countless site visits. My time was nearing a close, even

---

8   Liquidated damages are when a penalty of a high amount, sometimes in the tens of thousands of dollars, is imposed on a construction for not being completed on time.

with my extended internship. My last day came fast in August 2016, with a few weeks left of the summer before I started my final year of construction management. That Friday morning, I went around giving folks hugs and getting sweet goodbye gifts and plenty of good-luck handshakes.

Concrete truck chute washout

I got in the car and headed back home after three months of working with some incredible and inspiring folks. I was allowed to be curious and engage with a ton of questions. When I raised my hand and said I wanted to try something new, I was accepted with a yes. I drove away crying over the relationships and bonds I'd created and the safe space I'd had to learn and grow. There had been too few of these instances in my life, let alone when I was an adult.

Confident that I had the tools to get the job done and do it well, I was ready for what was next

SUMMER JOURNEY

Before my final year at Ball State University started and the next big chapter of my life began, I decided to take a trip out west to Los Angeles.

I had been working with a new friend who hosted a podcast about construction. He and I had a shared passion for the trades and were excited to bring visibility to them. After some exchanges and discussions during the summer, we made plans and agreed on a time for me to visit.

With a very tiny budget and a rental car, I made the drive out to Los Angeles right after the end of my internship with BSA LifeStructures.

This was an important milestone for me in many ways. I had never been able to afford to travel much outside Indiana. As a young adult, my finances had always been restricted to buying groceries and paying rent. Anything extra was not feasible. Now, for the first time, I was going to explore another area of the US and see the world with a much wider lens. There was a part of me that felt that if I didn't do this now, before I started a full-time role after graduation, there might not be another opportunity.

It took me a total of two long days to drive from Indiana to Los Angeles, California. I was fascinated with the change in landscape, the solitude different parts of the region provided, and the somewhat daunting, dark nights alone on the road.

CONSTRUCTION MANAGEMENT

In Texas, I saw my first wildcat. In New Mexico, I felt what it truly means to be in the dark. And Nevada amazed me with the Grand Canyon.

Arriving in Los Angeles late on day two, I made it to the suburban area of Torrance, where my fellow construction nerd and buddy was standing under a light post, waiting for me. What an exhausted but elated feeling to have made it as a woman traveling alone.

Over the next week, I took time by myself to explore. I sat on Huntington Beach, danced with the cold waves, and tried out new food spots. My little white rental Toyota whirred me all over, from downtown LA to the iconic Getty Art Museum. I buzzed around El Segundo Beach and checked out the nearby art shops and small stores.

Malibu was breathtaking, but the price tags on those homes made my jaw drop. I didn't get to see Keanu Reeves surfing, but I hoped for the possibility. Each day, I spent time on winding roads, visiting Hollywood and Beverly Hills. At one point, I even had dinner at Laguna Beach.

Time went by quickly, increasingly fueling my curiosity about the world. I felt so energized throughout my adventure, as I was realizing that there is so much more out there. Possibilities really could be what I made them.

Finally, I said goodbyes to my host and started the trek back home to Indiana. When I got there, I felt different. More encouraged. There was something so pivotal in taking the risk of traveling alone and gaining a new understanding of the world around me.

GRADUATION

My final year of college at Ball State came and went. I had wrapped up many of the harder classes the year before and was in the homestretch. As graduation inched closer, I started to interview with great enthusiasm. The three most prominent experiences included Talon Builders, McClane Solutions, and Champion Mechanical.

Talon Builders was a favorable option since it was recruiting for a position out in Northern California. With a bright promise of being placed in San Francisco or Oakland, I eagerly hopped on a plane. Once I landed, I quickly saw my assumptions start to dissipate. Oakland was a dirty place with trash in the water and all over the streets. The number of homeless people astounded me. I was shocked and disappointed but eager to keep my rose-tinted glasses on, as I continued to be optimistic about the interviews ahead.

My flight got me in late morning, and most of the scheduled events for potential candidates weren't until later in the afternoon. So I took my time walking around, familiarizing myself with the surroundings, and enjoying the adventure.

A rushed series of site tours and meet and greets developed an exciting rush of what could be a promising company to work for.

However, by the end of dinner, red flags started to appear. I'd showed up relatively comfortable in jeans and a hoody. I wanted to make sure my appearance was approachable and relaxed but put together. This mattered, especially as a woman, because being too dressy or showy could potentially invite the wrong male attention. In the same way, more relaxed apparel would encourage the predominately male company to see me in a friendly light, as someone they could shoot the gab with. Additionally, I knew if I was comfortable, I would come off as confident, which was important for making a positive first impression.

At the end of dinner, two employees were very insistent on me going out with them for drinks. After many courteous declines, I finally had to firmly say no and return to my room. In my mind, it was not professional to get intoxicated with coworkers. Plus, I wanted to be in the best form for the rapid interview sessions in the morning.

Jessi Interviewing for graduate position

The next day confirmed my suspicions.

At my second-to-last interview, the same manager who had pressed me to go out drinking was drunk. To this day, I can still see this individual's face in my mind. Yes, he reeked of alcohol and had bloodshot eyes from the night before. Immediately, I felt disrespected and incredibly disappointed at the lack of professionalism. I couldn't get out of that place fast enough.

Upon my return to Indiana, I reflected further on the experience. Within a few days, I was given an offer of employment, but I couldn't in good conscience work for a place that allowed their male employees to act so inappropriately. After careful consideration, I reached out to the HR director who had recruited me to explain the situation. I described my reasoning in detail. I wish I could say this was a one-off occurrence, but similar ordeals ended up repeating themselves down the line.

The next big interview opportunity was with McClane Solutions in St. Louis, Missouri. During my internship at BSA LifeStructures, I'd worked with a few folks from this general contractor. They were level-headed and encouraging people to be around. I was optimistic that this might be a great fit. After a very long and *early* morning of catching multiple flights, I was brought into a familiar setup of rapid rounds of interviews with a mix of different managers.

One after another, I was put in front of older White males who asked me a range of questions about my experiences. However, I realized that these men were all trying to gauge my level of ambition. Apparently, I was "too ambitious" when I mentioned pursuing a Juris Doctor[9] further down the line to expand my industry knowledge. I had been considering the next path after graduation, thinking about how I could elevate my current construction ambitions. Much of my coursework suggested litigation as an ongoing risk in the industry. A JD was a way for me to become even more of an asset to any company I worked for while broadening my construction knowledge.

Many of the questions I was peppered with directly involved my desire to get married or have children. This made me incredibly uncomfortable and didn't seem relevant to the conversation. After an intense day of questioning, I finally made my way back to the airport and home. It was yet another disappointing experience. Not only did I get asked questions by people I couldn't see myself reflected in, but I was put on the spot with what I considered inappropriate and gender-specific inquiries. Had I been a male, the questioning would have been much different. This is a fact I've observed over the years and have had my male counterparts confirm.

For my last big interview experience, I flew out to Minneapolis in the fall before graduation in 2017. My favorite professor, Mr. Mezo, had encouraged me to talk with a Kinesis Contracting recruiter about renewable energy initiatives. This brought me out to Minnesota in the vibrant fall. What was

---

9   Juris Doctor, or doctor of law, is a postgraduate diploma conferred upon the completion of all education required to practice law.

supposed to be a one-day interview session turned into two. Yet again, it consisted of the same cycle of older White males inquiring about previous experiences and expressing concern over how ambitious I was. This led to me being pulled into the HR office and being bluntly told that IB Construction didn't think I was a good fit. The recruiter I was working with talked to the human resources manager for some time and was able to get me in line for interviews the next day with Champion Mechanical, another subsidiary of Kinesis Contracting.

I was yet again rushed through a series of interviews with *more men* who asked what I wanted to do and what roles I saw myself in at the company.

The last interview of the day was one of the most memorable. I sat at a table with four men, one of whom dominated the conversation. The older of the four, Richie, kept poking holes in every single one of my answers. It was as if nothing I could say was good enough. When I was asked where I saw myself in five years, my response was that I wanted to take night school or weekend classes at some point to pursue my JD.

The look on Richie's face showed that he wasn't thrilled with this response. He kept asking about my interest in pursuing a Juris Doctor in the future. I explained in great detail how this would be beneficial to my employer and be a personal endeavor that I pursued at night and on my own time.

To say I walked away unsure about my future and a little confused is the best way to summarize the experience.

CONSTRUCTION MANAGEMENT

After a long deliberation, my recruiter finally walked out of the conference room, ready to take me to the airport. We started the mad dash significantly close to the time of my flight departure.

On the way there, he didn't have the greatest news or words of encouragement. He told me I interviewed well and that most of the panel saw me as a good fit. However, someone (he didn't name names) expressed significant concern over my postgrad endeavors. If I had to take a guess, I'd say Richie was the main character who was potentially going to foil my employment opportunity with Champion Mechanical. I was too ambitious in seeking a Juris Doctor, and this led to the false idea of me being a flight risk after a year or two of employment.

I was stunned and didn't have a lot to say that didn't feel defensive. At this point, as I was sitting in the car racing to the airport, I felt my gut sink. I was going to graduate without a job lined up. The recruiter and I continued to discuss the dilemma, with me adding details about what I'd said during the interview, how I'd explained why I wanted the JD, and how this was a benefit to the company. How could they not see me getting an advanced degree at some point as a great asset?

Ultimately, the recruiter told me on that drive that he had talked them down and was having a follow-up conversation with them in a few days. He advised me to hang tight and not give up hope.

By the time I got to the airport, I was frustrated. What had I said or done that was so wrong or different from the men around me? What was so bad about expressing a higher level of ambition—especially as it related to having a more robust toolbox to help the business and industry I was in? I felt like I had been smacked in the face and dismissed. Somehow, the incredible work I'd done and the fortitude I'd shown by bootstrapping myself from a welder to a construction manager was lost on most of these companies.

As I ran to catch my flight, barely making it through the gate before they closed the door, I wanted to cry. Sitting in my seat, I felt like I had chosen my moral compass over settling—and that was now going to cost me having a firm offer in hand before graduation.

Talon Builders wanted me, McClane Solutions ghosted me, and Kinesis Contracting said, "Nah, you are too ambitious."

Where was I to go from here?

CHAPTER 4

CALIFORNIA

I was in a new city with no one around.
What do I do now? I wondered.
Was this the right decision?

After a series of long trials of interviews in Minnesota and some significantly painful, anxious days of waiting, I finally got the offer from Champion Mechanical to relocate to the Twin Cities as a field engineer. I still remember the day I was sitting in the construction lab directly across from one of my classmates, Saleem, as I wrote the offer details on a Post-it and tried to quietly scream excitement while on the phone. In that moment, I knew my life was going to change drastically.

THE TRUTH ABOUT BEING A WOMAN IN CONSTRUCTION

A few months later, as I was getting ready to graduate, I scheduled the movers and found a two-bedroom apartment in the West Metro area of Minneapolis.

My last few days working at Red Lobster were bittersweet. Over the previous two years, shortly after I'd moved up to Muncie, Indiana, for my undergrad at Ball State University, I started serving to cover many of my bills. Waiting tables at Red Lobster was flexible enough to accommodate my classes. In hindsight, it also gave me new skill sets and a crash course on dealing with difficult people. I had learned how to smile and keep calm, even when I wanted to smack someone for an inappropriate comment or the thirty cents they left behind as a tip. As I prepared to leave that job, people I despised or found annoying were excited for me and gave me little gifts and congratulations. I, to this day, still have the jam-packed grad card full of goodbyes and good wishes.

Jessi in Red Lobster Uniform   *Muci in Grad cap*

CALIFORNIA

Jessi in superwoman pose

In May 2017, I graduated and made the journey to the Twin Cities in a U-Haul with my fur babies in tow. My mom followed in her minivan, packed to the brim.

Suki, Muci, and Beasley were part of my original pack of four cats and two dogs from the previous four-plus years of my life.

Muci was a sweet, bull-boxer mix. She was mostly white with brindle spots. Weighing fifty pounds, she'd shown up one day on my doorstep, scared, starved, and abused. I took her in and named her Princess Muci Moo.

THE TRUTH ABOUT BEING A WOMAN IN CONSTRUCTION

Beasley, a tiny black runt of a cat with stark green eyes and a mini white mustache, was a rescue from an old roommate when I was working at Walmart. She was ignored and malnourished, so I took her with me when I moved out.

Suki, my newest addition, was an all-black medium-size whippet-terrier mix. Her narrow hindquarters and her ears, which would pin down when she ran, made her the perfect companion for my aging Muci.

Suki and Muci snuggling

My mom was a nervous wreck over this move. We had several hard discussions leading up to my decision to relocate. I knew that if I left her behind in Muncie, she would be isolated.

Without the right support, she could have developed further mental and physical limitations. So she decided to come with me. Financially, I was stepping into a much better place, which offered stability to both of us. Even though it was hard, and I could tell she was scared (because she'd never been to Minnesota), she also didn't want to be left alone in Indiana.

With many whimpers and meows in the back of the U-Haul, we made the ten-hour-plus drive. It was liberating to see the landscape change. As we edged closer to Minneapolis, chowing down on cheese bites and beef jerky, the city view became clear. When we arrived in the fast-paced traffic of a new state, the sense of a great journey vibrated through my body. I cranked up the radio as I navigated the traffic of the busy metro until we finally arrived at our new home.

Nothing could prepare me for what was to come next.

My first day at Champion Mechanical was a series of onboarding sessions, including a tour of our floor of the building nestled in Eden Prairie and introductions to folks. The small team I was brought into was primarily made up of fellow recent undergrads. Many of us were starting as office engineers or field engineers. Typically, this is an entry-level point for individuals working their way up to become a project manager or superintendent. I'd been brought in as an office engineer. I would be helping while learning how to estimate and support bid reviews. There wasn't any real transparency as to what my responsibilities were going to be. Instead, I was going to learn the business and how we got and built work at a high level.

Two of the new peers I was onboarding with—one of whom happened to be a fellow alumnus of my program at Ball State—were told within a few hours of arriving that they were going to be sent out to a project in California without any indication of when they'd return. There was lots of nervous, anxious energy as both individuals cycled through a spectrum of responses mixed with surprise. One was frustrated, and the other quiet.

There was little to no detail given on the project these new team members were about to embark on. Internally, I was cartwheeling with relief. I didn't want to repeat the intensive ordeal of packing my mom and myself up to relocate again. My school buddy ended up accepting the assignment, even though our shared truck ride home suggested that his anxiety was a mix of frustration and excitement. The other newly minted team member said he didn't wish to go out to California and was immediately fired. On the spot. Terminated. Just like that. As though he were nothing more than a goldfish's two-second memory.

This should have been a foreshadowing of what was to come, but I didn't give it any more thought than I would give a rare occurrence. The week continued with the wide variety of new team members participating in a series of team-building exercises, including a basketball game using bumper cars and beer (a.k.a. Whirlyball). The weeks carried on as I started to find a rhythm in my schedule by joining a gym, getting up early, and learning from my new colleagues. I started to gain perspective on the tiny network I was building. Future golf outings with some of my office mates were being planned, and team dinners with laughter and rising confidence all suggested this was a good place to start.

Then I started to notice an increasing number of calls that came in for a few projects from Northern and Southern California. Even though no one explicitly warned me or suggested I would end up on a project, I started to get a gut feeling.

Roughly around the one-month mark, I started to investigate buying a car. Up until this point, I was ridesharing or using my mom's Honda Odyssey. I ended up purchasing a black 2012 Dodge Journey. Then, within a few weeks, I got called into my manager's office.

She told me she had an exciting assignment for me, that I was going out to Northern California to work on a refinery. In the moment, I was hesitant. I was still trying to figure out Minnesota and get my mom settled. I told my manager as such, and she assured me that my mom would love it out there. I didn't agree but didn't dare say so out loud. My mom had already had a hard enough time moving up to the Twin Cities, and now I was about to uproot her yet again. I had been taking care of her for the last three years. She had mental and physical limitations that came with age. Moving her would cause instability, and she needed to feel grounded and safe. Repeating the process so soon would have been detrimental to her health, not to mention I had no idea how accessible Northern California was going to be. After some time, many discussions, and a quick reflection on the team member I'd seen get fired in the snap of a finger, I called my mom and said, "I am going to California."

We discussed it for some time, and both agreed the best choice was for her to stay put in Minnesota. Soon, there was lots of scrambling to find an affordable place near San Francisco, arrange yet another U-Haul, and pack up my belongings.

Early one morning in the late summer of 2017, I hugged my mom goodbye and said, "See you later," to my eldest pupper, Muci. With my youngest companion, Suki, in the crate behind my driver's seat, we started our trek west.

The emotions I felt were overwhelming. I had big tears in my eyes and was scared to *death* of what I was doing and what I was leaving behind. After some time, I let the music carry me as I let it be a distraction while I watched the dark morning transition to a peach sunrise. I had made the drive out west to LA the summer before with the hope I would be returning to some sense of familiarity. Nothing had prepared me for the mix of fear and excitement I held in my chest as I drove toward the unknown.

As I started to consider the new challenges ahead, there was much that I was not prepared for. After three long days of driving, I was excited, scared, and ready for the next chapter. When **I stepped into the tiny studio on my first night in California** after the two-thousand-mile drive from Minneapolis to Point Richmond, California, an overwhelming feeling occurred.

I was truly alone in a strange city thousands of miles away from any form of comfort.

CALIFORNIA

I remember the intimidating darkness surrounding me when I was laying out my yoga mat on a makeshift wooden platform left by the previous tenant. There were so many different emotions, but the one that surfaced the most was fear. I had never been so scared in my life—not even when I was turning eighteen and moving out of my mom's house.

There was no sense of jubilation or accomplishment. Just a bone-chilling "Oh shit, I am alone" in a new city, with no connections, in a new studio I had put a deposit on sight unseen.

Right after I got there, I was walking down the street with Suki and talking on the phone with my mom. As I spoke, I observed the tiny dead squares of grass and looming San Francisco—cut up into smaller apartment-style houses. The physical push needed just to walk up the hill to find a place for Suki to potty was immense. Not only had I been there for only a few hours and seen only my landlady, but there was no one I could refer to for comfort or familiarity. No peers (that I knew at the time) were close by to chat up for information about the neighborhood or offer warmth to my new surroundings. I will never forget my first night in Point Richmond in the beat-up home on Tunnel Avenue, where I slept, wary of those outside the walls of my new sanctuary.

After a few nights of getting settled, taking in all that was around me, and observing the different environment, personas, and mode of life, I started to feel my anxieties slightly fade. Eventually, things started to settle when my U-Haul pod arrived a week later and a new bed replaced the Target blow-up mattress.

Over the months that ensued, I became more familiar with my residence. This included the odd hours of night and morning when the dark energies emerged. I typically had to be up early, when it was still dark out. On the weekends, I would stay up late at the corner rugby bar, which was well-positioned in the center of the busier area where I lived. I would observe the homeless in odd pockets, spread out hunting through garbage or sleeping on benches while other folks wandered in drug-induced dazes.

Even now, the feeling of what I experienced still feels foreign in hindsight. But I am grateful for the path and challenges I faced. As Chelsea Dinen says so eloquently, "And that's when the world becomes your oyster. Moving cross country alone taught me more about how independent, strong, and capable I am than any other experience."[10] **This jump was one hell of a sacrifice I have not forgotten.**

REFINERY

The first day on site was a nice, easy warm-up. I quickly made my way to the training trailers that were outside the refinery where I was going to be working. All new employees were required to attend a three-day training on refinery safety basics. Much of it was similar to the OSHA training I had been through before, while other parts were a bit of a shock—including the fact that we had to walk around in bright-orange jumpsuits

---

10 Dinen, C. (2018, September 4). *Why being scared to relocate isn't such a bad thing.* https://chelseadinen.com/scared-to-relocate/

called Nomex.[11] These were supposedly major protections in case the plant itself caught fire or an individual was caught in a scenario where a fire might occur.

Women on day shift in Nomex

The training was relatively straightforward, other than the warnings about how to prepare for being blown up because it was a refinery. Hopeful to learn more about what I was going to be doing on this project and exactly how my skills would come

---

11 Nomex is a bright-orange jumpsuit made of fire-retardant material that prevents the individual from going up in a blaze. This is essential personal protective equipment in a refinery.

into play remained a large question. Before I left Minnesota, I was told that I would be in charge of post-weld and preheat treatment operations. I made sure to keep the booklet I was offered, which provided some context of what those unfamiliar operations were. Grateful for some of the tricky, put-on-the-spot moments I'd had as a kid while working at my father's contracting business, I was confident I would learn the role quickly.

I met another field engineer who had come down from Canada and was anxious about his new role as well. Neither of us had been offered much detail or help in getting a place to rent. We quickly befriended each other while we climbed practice rungs with our new harnesses.

When I was in high school, I'd daydreamed about moving out to California. As a kid who never fit in, I envisioned the West Coast as my true home. I had a desire to find like-minded souls who would appreciate each other's oddity and respect the differences humans embody. Once I was living in California, I started to be hopeful that my new surroundings would embrace my teenage daydream as a reality. But what started as a potential sweet new beginning post-education on the other side of the country became one of my biggest living nightmares.

I soon met my bosses, Eli and Bowen, who were titled project engineers (commonly known as project managers in other companies). Both had worked on different projects for Kinesis Contracting for a handful of years. Their kind words of encouragement led me to believe I was in good hands with capable leaders. One of the few introductory site walks I received offered

CALIFORNIA

an explanation of how the site was broken up. I received very little work product or training on what post-weld heat treatment (PWHT)[12] or preheat weld heat treatment (PRHT)[13] was in the refinery. I was given a roughly assembled plywood desk in the bright-red bomb box[14] and a tablet to conduct my business.

After meeting with the team I was working with (minus a superintendent), I went to work figuring out my team's needs and trying to rapidly understand my new role. There was no clear context or communication about what my role and responsibilities were. There was nothing on paper—just a loose verbal statement of, "We are counting on you to turn this around." There was all this expectation but no clear directive.

After a month or so, a superintendent from Alabama showed up. I gave him a quick tutorial on some things I had discovered and built. Tools I had created included work packs that showed the pipeline numbers. I was eventually able to upgrade these with tentative Revit color printouts that indicated location, size, and even angle of the pipe.

---

12   Post-weld heat treatment occurs after a pipe has been welded to help with the expansion of the gas or fluid line. Typically, little things that look like pink chiclets are wrapped around the pipe and turn red when hooked up to an electrical current.

13   Preheat weld, similar to post-weld, occurs prior to the pipe being welded. Chiclets are set up around a weld or seam opening, then heated up when the welder fuses the gap together.

14   A bomb box is, in essence, a large stack of what appears to be a shipping container. It protects people during a refinery explosion.

The idea was to optimize and encourage the PWHT and PRHT teams I managed with clear direction in a jungle of pipes, materials, and other work that went on nonstop twenty-four hours a day. Another thing I brought to the superintendent's attention was a ticketing system to improve communication from the various pipe fitter[15] foremen who need the PWHT or PRHT process done so they could get in a queue to prioritize and simplify the work for our teams.

Together, the superintendent and I seemed like a great match. But after a few weeks, that hope started to crack, and my life turned miserable.

First, it was small things, like comments from him about how great I looked in Nomex or why my lips pouted when I was putting work packs together. This escalated into late-night texts and videos sent to my work phone of him playing guitar while drunk. I did my best to ignore and even downplay it, but I was also worried that if I didn't respond, it would make the next day worse.

This went on for months and gradually got incredibly hard to deal with. Finally, I got to the point where I would avoid him all day and even got my crew to help me set up a headache shack[16] (a.k.a. a large metal structure that houses a simple desk-

---

15 The pipe fitter preps, installs, and repairs a gas or fluid line. Apprentices will fit up or line the two pieces of pipe while the welder fuses the seam or gap closed.

16 Headache shack, or mobile digital plan station *Jobsite Boxes*. KNAACK. (n.d.). https://www.knaack.com/products/jobsite-storage-solution/jobsite-boxes/FieldStationSeries/118-01

CALIFORNIA

like structure that can close when not in use, also known as a mobile digital plan station). My crew coined "Cuppycake" as a nickname for me. (No idea why, but it was cute.) They set me up so I could work outside the bomb box more to be near them and closer to the action. In my mind, this made perfect sense. Less time going back and forth to the bomb box meant less potential interaction with my superintendent.

Around October 2017, the Bay Area had one of the worst wildfires in history when Sonoma wine country, which was directly across from the refinery, caught on fire. A gender reveal had gone bad, catching the winery and many surrounding areas ablaze within a few hours. It was a devastating event that many of us were not prepared for. For about three weeks, we tried to operate in a haze of smoke that didn't let us see much more than a few feet at a time. Our safety engineers did their best to make sure folks had the right PPE[17] and masks.

Unfortunately for me, I hadn't been fit-tested, so I couldn't utilize one of the more heavy-duty masks[18] that would filter the heavy smoke. This was a significant hardship that made many of us feel ill or unstable when we worked on-site. Most of the time

---

17 PPE stands for personal protective equipment. It typically includes safety eyewear, gloves, a vest or Nomex, and steel-toe or composite boots. These are the basics to keep yourself safe on a construction site.

18 Heavy-duty masks can range based on function. In this instance, they were a full-face covering with two air filters that stuck out. Think postapocalyptic movie, and you'll get the idea.

when I was outside, I had a bandana folded over and tied behind my head to help cover my mouth and nose. Safety glasses kept my eyes from welling up or getting significantly irritated. Even then, it was still hard to breathe, which meant I was stuck inside the bomb box most of the time. What a scary time for most of us, who were charged with field crews and operations. Many of us thought it was insane to work in these conditions, which placed significantly increased risk on our people. However, the higher-ups didn't want to further delay the project, which was already behind schedule.

At the same time, things had escalated with Eli and Bowen, the site project engineers. Both had started to target me in the work I was doing. At times, I would get called into the office and asked a bunch of questions about the work being done. Each time, either together or solo, they would test me to see if I had a clue about what was going on. They started commenting on my work product, questioning what I was doing, why it took so long, and why there was so much repeat work.

I constantly felt on the spot, especially when one of them would take an interest in my work and show up while I was walking the site or checking in with my crews. Where was this interest when I started? Why were either of them suddenly so interested in what I was accomplishing with my crews? Any mistake was an opportunity for them to headbutt me with "You should have known better" and "What is the lesson here?" This was a new layer to my added stress on the project. Now I felt like the eggshells I was walking on had sharpened to a fine point. The only way to survive was to avoid them altogether or be hyper-prepared. To say stress and cortisol levels were high was no joke.

All of these things continued for about six months. During this time, I had befriended an older structural steel engineer, Olga, who was a total badass with a "don't mess with me" attitude. She and I would walk the job site, make safety corrections, talk with crews, and share updates on each other's work.

Field sisters at my headache shack

One time, we had a significant safety incident when welders were working on top of one of the metal bridges between units. They were getting sparks all over, which were falling onto and burning everyone working below. She and I were at the lowest level and yelled up without any luck. She took off to the top while I worked on getting the others below moved and radioed the safety engineer.

Once everyone was settled and the welders above were safely resettled with the proper spark mitigation, Olga and I practiced yelling. She had a way of yelling deeply, whereas I sounded like a baby bird starting to choke at a high pitch. Olga was great at explaining how to breathe in deep and then bellow out with my stomach. She was such an incredible person and mentor. Our time together eased the hardship we both faced on that grueling job.

The project continued to get worse, with new facts being uncovered. For instance, our onsite team was offered new details on the project from higher-level leadership. Whoever owned and operated the refinery prior had worked with a completely different contractor ten years before passing off a 40%-complete project. Instead, we were discovering the refinery was closer to 15% built. Crews were getting more frustrated and tired of the redirection. Materials were constantly being delayed or lost. Tradespeople were conflicting with each other in areas. It was hard to work with and attempt to keep teams at ease or continually productive. It was a two-person job.

My role was to primarily be focused on providing information and resources for the crews. The superintendent's job is to keep everyone moving and make sure conflicts or problems get resolved. In a perfect world, the superintendent and the field engineer work in harmony to support the operations they are responsible for *together*. I was still the go-to person when anyone on my crew needed help. They were supposed to talk to me instead of reaching out to the assigned superintendent. He would hide out all day in the TEAM[19] tent while I was out there working my butt off, getting generators in place, checking wraps, verifying lines, or climbing up to answer a question.

Between the inappropriate attention I was getting at night from my do-nothing superintendent and the increased hostility I faced from the project engineers, I was feeling unfit for this role. I started to doubt myself and my abilities, and I internalized the comments and micromanagement offered by Eli and Bowen, who were undercutting every effort I made to keep my scope well-managed. I was still trying to make this job work, but I was struggling and feeling more defeated by the day.

To my surprise, after a series of empty threats about leaving, my superintendent left and never came back. I have no idea what happened or if he found a better gig, but I was relieved. His departure, however, also made me feel alone and isolated. Yet again, I was back to being the field engineer and superintendent.

By this point, we all were working thirteen hours a day, six days a week.

---

19 TEAM is a specific subcontractor focused on managing different types of data. At the refinery, two dedicated individuals monitored the electrical feedback from the Post Weld and Preheat electrical lines when they were turned on and cooking a pipe. *Integrity management: Asset integrity for pipeline & aviation.* TEAM, Inc. (2024, May 21). https://www.teaminc.com/

Talk about being exhausted and worn out. As a woman on a site that was 95 percent male, I had to deal with dismissals about direction and safety procedures. I even got grief for telling workers not to smoke on the job. Having a period on-site was miserable because I had to strip off an orange Nomex and sit inside a porta-john.

I had no life and was feeling more isolated by my managers. I was stuck. At the time, I was paying for my tiny studio in Point Richmond and my apartment back in Minneapolis, where my mom was.

Over the next four months, things got worse. To mark my eight-month anniversary, I was placed on a personal improvement plan (PIP)[20] for my "poor performance" and still left alone without a superintendent. There was no real reasoning provided for the PIP. Most of what was given initially, not in writing, was that I had a hard time connecting with the team. Eli and Bowen outlined my behavior as the problem. They claimed that my interactions were too direct and I didn't smile enough. This all felt very directed at my not being "soft" or "warm," as though I didn't fit a stereotype they clung to.

Ultimately, I was being punished for my more "masculine" approach to working as a confident and direct woman. My PIP felt like an attack on me being a female who took charge

---

20 A PIP is typically outlined to manage a specific focus of professional development. Also used as a last resort to document poor behavior or an underperforming employee before they are fired.

and got stuff done instead of wringing my hands and asking for permission. There were some small comments on the work I managed not being done well, but there was no tangible evidence. And nothing gave credible direction on how I could improve or how both the project managers would help support me.

I was managing two shifts of PWHT teams by myself after being on the project for nearly a year. I was now expected to stand up in our daily morning meetings and give updates. The superintendent would call me asking for help on their problems, such as "Your generator is in the way" or "Your lines for the wraps are blocking access." And there was no reprieve from the other superintendents. They gave no shits that I was a field engineer doing the job of a superintendent. Instead, they grew more and more frustrated that I wouldn't get things done fast enough, or they wanted to know who made a final decision on work being done.

I was feeling more helpless and exhausted at my job, and the self-doubt started to weigh me down. I was becoming more unsure of my profession and needed time to reset. Luckily, fate was on my side and gave me a break by sending me out to Denver, Colorado, for a weeklong company job training session.

CHAPTER 5

DENVER

Exhaustion feeds negative self-talk. Once that cycle starts, other things follow, such as an unhealthy diet, poor exercise habits, and loss of sleep, all leading to burnout. You are priority number one and need to pay attention to the early warning signs. Sometimes, the best decision is to get out!

After about eight months into the refinery project, I left for a weeklong training opportunity in Denver, Colorado. This became a fortuitous event and a much-needed break from the refinery project hell I was living in.

It was my first time in Denver, and it was a mini adventure. Each day, there was focused training dedicated to over a hundred young or entry-level Kinesis Contracting field employees. Most of us had very similar titles—either field or project engineer—

which was a way to tier out construction managers. Field engineers went all the way to Field Engineer III, and project engineers progressed very similarly. A project engineer was the closest thing we had to a project manager, whereas a field engineer was a blend of entry and assistant project manager. It wasn't the clearest ranking, but that's how the company operated.

On the first day, we went into some training around bolt connections and framing. Split into a series of small teams, we problem-solved the different construction challenges. It was a great experience and good team-building too. Day two was my favorite because we had a welding course. I was thrilled, since this was my first career calling. During the day, I ended up assisting others in explaining processes and techniques.

One other woman was there, and her name was Amber. She was driven to weld perfectly and turn in the best possible work product. With my hood down, I worked with her while I stood closely behind. I offered some setting adjustments on the welder and, at times, told her how to angle the welding gun.

She and I hit it off and ended up in a series of conversations about the ongoing challenges we both were experiencing with the different branches of the company. We left the training after exchanging information, and we even found ourselves commenting on or supporting each other at future sessions during our time in Denver.

By the end of the week, I was feeling a little more motivated to get back to San Francisco. I was even optimistic that there was light at the end of the tunnel with this company.

That didn't last very long.

When I was at the airport waiting for my flight, I got a phone call. It turned out that I was going back to the project, but I'd be in a new role. I was switched to nights as an Area Lead Pipe Fitter Engineer for four different sections of the plant. There was no discussion or inquiry about whether I was okay with the transition. Strangely enough, this was like a promotion (with more work and slightly more pay). Still on a PIP, I came back with a new superintendent who had been imported from Canada. He and I hit it off immediately.

100 feet up at night

Upon hearing this news and experiencing this change in environment, I made sure to go to work on pulling together every document, process created, work plan, and file. I was no longer in charge of the Post Weld or Preheat team. That job had been given to a twentysomething young man who'd just recently graduated. To this day, I wish I had before and after photos of this role.

On day one, I'd been brought to an empty wooden plywood cubby with nothing more than my work laptop and a notepad. When I left my role—and after "earning" a PIP—I'd assembled work plans that included Revit printouts (3D modeling) for pipe locations, pipeline numbers sorted in order, and color-coded documents showing what was completed, in progress, or yet to be done. This new kid they'd brought in to take over was barely twenty-one (I was twenty-eight at the time), and he had a full stack of drawings, file drives, and processes built and laid out on his desk to help him adjust on his first day.

I'd started with *nothing* and built that into *something*.

DRY PROMOTION

I went from managing four different crews (two during the day and two at night) for PWHT/PRHT to working nights. What a difference it made. I hated the endless days and nights of feeling exhausted. I gained weight like crazy and felt like a vampire on Sundays—the one day I had off each week. This is all to say that I so enjoyed the quiet of the night on the refinery. I quickly gained favor with my crews, who seemed to adore my collaborative approach to problem-solving and my empathy and consideration of their needs and wants. There was much to learn in this new role, with a slow bubble of comfort, as well, among the chaos.

My superintendent from Canada, Lucas—whom I ended up nicknaming "Frenchie" for his thick French-Canadian accent—was a refreshing change of pace. He was kind and warm but also very direct. He and I went to work managing several crews, coordinating between the two of us to make sure they had what they needed, their questions were answered, and work progressed efficiently. Together, we problem-solved, with him being the enforcer and me the resource builder.

It was a special partnership. We just got each other and knew how to back one another. After a few months passed, work was moving as best as it could among many confused trades, materials were scattered in a borrowed parking lot a mile down the road, and crews came and went as they pleased.

I started to find some women on the site (the few of us who existed), and we began building our little ecosystem. Together, we had quite the range of skills. There was a newly hired field engineer, Trinity; Jett, who was a laborer;[21] and Tricia, a pipe fitter apprentice. There was also the safety engineer, Taylor, with whom I worked closely on days after he was moved over to

---

21 Laborers are the unsung heroes of any job site. They run the gambit of work performed, including picking up trash, cleaning, and unloading and loading tools or materials.

nights. Each shift, we would find opportunities to check in with each other and listen to the other vent. There were days when we would bring goodies in for each other, it being a safe space where we could learn from each other and feel supported.

Unfortunately, regardless of our best efforts, both the other field engineer, Trinity, and I started to struggle with the crews and some of the on-site leadership. She was Black and was treated horribly. Sometimes, name-calling was involved. At other times, expectations were placed on her without any clear direction or communication. The impression I gathered was she was getting set up to fail. Her superintendent didn't like her, either because she was Black, because she was a woman, or because she was heavier than a lot of the women on the job site. None of these were good reasons to push her out.

Many nights, after we got the crews set up, we would walk around, sometimes sharing the load of inspections with our prints or checking out new work that had started up in an area with other trades. Other nights, we would just walk, avoiding the bomb box, where she or I might have to deal with certain managers who were still on-site. There were so many nights when Trinity just broke down in tears because she was so frustrated and unsure about what to do. I was lucky with my superintendent Frenchie. He did his best to look out for himself and me. He was and still is rare among men. Maybe it was the Canadian in him. Who knows, but he was a shining light in the dark night of the storm.

During this period, I had also become closer to one of the

safety engineers, Taylor. He and I had met while we were both working on the day shift. We built camaraderie around the events and chaos going on at the refinery. He would tell me stories about the other projects he had been on in different states. Together, we created a safe space for each other during the night shift. We would take late-evening site walks, working collaboratively to address some of the challenges observed on-site.

There were other good moments too. Checking in with most of my crews typically led to a brownie bite or shared French macarons from some kind soul who'd brought them in to share. I always tried my best to stay positive with the teams doing the on-site work. We'd swap stories about the weekends or share our favorite vacation memories. Together, we would laugh, hearing how someone got stung for the first time by a jellyfish or learned to surf and failed. The camaraderie was real. In some ways, we were like a dysfunctional family—a ragtag mix of pipe fitters, electricians, and crane operators.

My senior project engineer—Chad, another Canadian team member, who oversaw nights—continued to evaluate my PIP in the background. He would attempt to be helpful, even offering positive reinforcement at times. Late one evening, when we were sitting down, looking at my progress update for the PIP, he told me that I should try taking some emotional intelligence training. Even as I write this, I chuckle. What this really meant was that I needed to know my place as a woman, smile more, be less serious or focused, and remember that the men were always right. I was in trouble because I was forthright, and the project engineer on days, Eli, needed a scapegoat to blame when things were messed up.

Of course, I nodded and thanked Chad for his input, even if he was only a year or so older and out to lunch as to the real reason behind my PIP.

I was a wreck 98 percent of the time. My insides were always tied in knots. I was confused over what I was doing on the refinery and scared stuck. Literally, I didn't know what I was going to do next, who would hire me, and whether I should stay in Northern California. All these questions continued to flow through my mind. It's not like I was making a ton of money, and I still had my mom to care for back in Minnesota. The black hole was this job, and the support was barely enough to keep me in one piece.

Things went from okay to worse on the night shift in a matter of a few weeks. Karen, the mother of one of my foremen, was running another crew. She was a craggy woman who just hung out at her headache shack. She didn't do much, and neither did her son. It didn't matter, though. They'd been hired in as foremen with the union. Occasionally, he and I would have some disagreements. When he got angry, his bushy red mustache and face would go beet red, and his eyeballs would bulge. Things shifted poorly, though, once his mom, Karen, was taken off the job due to her crew's low productivity and quality of work. Many of her team members joined mine, adding to a higher count of people my foreman and I had to manage.

The increased size offered me the ability to train a secondary or backup foreman.

As one can imagine, the situation continued to degrade. The foreman continued to be more resistant to any communicated directions or information. Over time, I had to get my superintendent, Lucas (a.k.a. Frenchie), more involved, to the point where he would come to this one area's crew and be their dedicated resource or communicator. The foreman with the bushy red mustache would just flat out refuse to follow any direction from me, or he would start to yell and complain. In several instances, he shrugged off the direction for the night (coordinating work scopes with the day crew to avoid gaps or repeat work was a must), then cowboy repairs or new welds and neither document it nor have notes ready for the next day.

One night, I was walking all the shacks, checking on each team to see how they were doing, providing turnover details from the day crew, doing general check-ins, and so forth. The foreman with the bushy red mustache got *pissed* at me. I was relaying information and enforcing the need for certain steps to be taken to him and the crew for this part of the area. He stepped so close to me that I could see the red veins of his eyeballs. In a low voice, he told me, "This is my crew, and they don't take orders from you."

That was it for me—my final breaking point. It was clear as day. I'd been yelled at plenty by different foremen, superintendents, and even tradesfolk who got defensive when I asked them to take care of basic things, like putting a hard hat on. This was different. Internally, my body and mind *exploded* when the foreman got in my face and threatened me.

Without budging, stepping back, or even twitching, I turned my head right to the foreman in training and asked if he had any questions for me. He slowly shook his head no. Everyone was quiet. It was like the Earth stood still.

I calmly walked off and found my superintendent, Frenchie. Immediately, I blew up in a flush of anger—hands shaking, body on fire—and told him I was done. There was nothing else I could do to improve this situation, and I wasn't going to keep putting myself in harm's way.

I don't think I'd ever seen my superintendent so freaked out. He kept trying to calm me down and understand the situation. Then he went over alone and had some words with bushy red mustache.

Next thing I knew, we were all sitting down—the general foreman, Frenchie, the foreman in question, and me. I ran down the situation, and Frenchie reinforced it with the special accommodation we had made to deal with the escalated situation. This included me no longer making a stop at bushy red mustache's headache shack. If he had any questions or needed support, Frenchie would provide it. In essence, we were not to speak to or see one another.

A series of things happened next. Bushy red mustache took about a week off (whether it was scheduled or impromptu, I didn't know). At the end of his week off, I had PTO put in for a long weekend. For reference, the foreman's time off and mine gave us about nine days apart—time to cool down, reframe, and refocus.

When I came back from my time off, the general foreman verbally slammed me as soon as I walked in, saying I was harassing bushy red mustache. He listed out all the things I was doing to make bushy red mustache's life hard and said that I was overstepping in my role. Ultimately, according to him, I needed to know my place, and he assured me that the union was going to protect the foreman at all costs.

It became insane after this point. New rules were put in place saying I was no longer allowed to have any sort of communication with the foreman. And his crew was going to be reassigned to another engineer.

In case you didn't connect all those dots, let me help. I—the woman who brought out references, directives, and updates from the day crew, along with leadership and schedule changes at the start of every shift—was being thrown under the bus for "harassing" the foreman.

Remember, I wasn't the one who got in someone's face and acted threatening. He did. I, the victim, was being punished for reporting the issue after taking weeks of his verbal abuse and dismissive attitude about the work he needed to get done every night. The problem wasn't the job I was doing or even that I was the one doing the harassing. The problem was the fact that I was a woman who was giving orders.

I quit two months later after getting a sweet offer from another contractor. Not only did I get a 20 percent pay bump, but the opportunity was in Portland, Oregon, which offered a fresh start. The cherry on top was that two weeks before I accepted the offer, I was released from the PIP thanks to Frenchie, who aided the conversations around my development and ultimately saved me from being fired.

I quit *exactly* five days later after receiving the offer from my new company. On my last day, I sent out a two-page email to every part of Champion Mechanical's leadership, letting them know about the harassment and indecent abuse I'd faced on the job site.

I poured gasoline and lit the match, then watched that bridge burn.

CHAPTER 6

OREGON

"You are too direct in your email responses and how you talk to folks on site. How can you warmly address the problem without the individual [man] not feeling threatened by your ask or redirection?" Are you sure it's his fault and not yours?

$\mathcal{P}$ortland, Oregon, felt like a fresh start—like I had found a cool ocean breeze to wash away all the gross smog and pollution gathered from too many months surrounded by toxic individuals. I remember, clear as day, the first time I stepped into the Portland office. It had a cool vibe and was very woody, with bicycle racks filled with commuter bikes. What a drastic difference it was compared to the bright, breezy Northern California.

The night before the interview, I did all I could to prepare. I mentally practiced my answers in preparation for different types of questions. How would I respond? I even scoured my printed portfolio, which was proudly covered with an image of a group of hard-hatted women wearing bright-orange Nomex. I was looking for any reason they wouldn't hire me.

I was wrong, and my fears washed away about a week later. I'd made some sort of impression with my interview, because **they were ecstatic to bring me in, and fast.** This had been, indeed, one of the better interviews in my career so far. Questions were relevant to my job experience, and both managers were **impressed by my portfolio. They bumped my salary up** after some negotiations and included a moving stipend (as any good company should).

Pulling the offer letter off the printer at the bomb box while on site in Richmond, California, was one of the tensest, most exciting, and most anxiety-inducing moments. It felt like that single piece of paper was my only ticket out of the dark environment that I felt was consuming me in every waking moment—and even, at times, in my dreams. I was desperately holding on to it for dear life, like a lifeline.

But I didn't get to the printer just in time and ended up competing with one of the project managers for the growing stack of printed documents. He sifted through it and found my **offer letter, which caused me to blurt out, almost in a knee-jerk reaction, "Oh, that's mine!"** Then I scurried off to my small wooden plywood corner desk.

Later, I would approach that same project manager, who had been with the company for almost thirty years, to boldly say, "I did my best and gave it all I could."

He just stared at me, expressionless, without making any remark. It was as if I was already gone and the work I'd accomplished didn't matter.

As I've explained, I had just been released from my PIP when I turned in my notice. No one batted an eye, and, for the most part, people were congratulatory or altogether dismissive. The silence that I was able to soak in between my notice being submitted and my final day was bliss. There was a solitude in knowing I was mere days away from a new beginning—hopefully one with a better company.

My eighteen months of working with Champion Mechanical Contracting came to a close, and I started once again the process of relocating and coordinating with my partner at the time. I packed up my tiny San Francisco studio to make the nine-hour drive up north. It was a stressful adjustment, trying to find the best place to live that was close enough to the project I was going to work on in Beaverton, Oregon.

Fast-forward a few weeks, and I was meeting my new team for the first time. New home, new team, and new project. Instantly, I was blown away by the trailer I was working in. No longer did I have to wear bright-orange Nomex (a.k.a. a fireproof prison suit). I had an actual desk, not a plywood arrangement that a carpenter had thrown together. I had two brand-new, clean monitors, a new laptop, and indoor plumbing.

Let me repeat that one: *indoor* plumbing. If you are a woman, you know that porta-johns are a nightmare. They stink to high heaven, men do some disgusting things in there, and periods, in general, are not a blast. Add in having to squat dance and clean up in a porta-john, and it's no joke.

I thought I had landed the best job ever, working on interior scopes for Momentum Builders. Managing color selection and furniture scopes and scheduling interior designer meetings was an easy, almost pretty, job. This was a major project, with millions spent on the building's branded exterior gold paint alone. I was working with a field engineer who had just gotten out of college a few months before coming out to his very first project. His name was Breck, and we got along well. He and I swapped stories. With him being from Texas and me having grown up in Indiana, it made for some interesting comparisons.

Our project manager had a Boston-meets-New-York mashup of an accent. He was a super short man named James who walked like he couldn't get anywhere fast enough. Loud and overly blunt, he was the type that says things for attention because it'll turn a head or two. For the first few months, we got along. I appreciated the bluntness he offered, and in return, I was equally blunt. This was okay until he didn't like how straightforward I was.

One day, I was expressing to James my concern over a much older project engineer. When I say he was older, I mean that he was easily ten years from retirement. He was inattentive, hyper-focused on the wrong things, and had absolutely no experience in the field. He was a walking hazard who was going to get hurt—or worse, get someone else hurt. I was speaking candidly with James and expressing my concerns when he indicated that this project engineer in question was protected by the company. His age automatically prompted his higher status role even though his background was in engineering.

Toward the end of the conversation, I rolled my eyes, not at James but at the situation. He got pissed. He told me not to roll my eyes at him. And it wasn't in a "Hey, that's not cool" manner. It was more of a slap across the face. Like a "How dare you?"

Soon after this encounter, I started to find it harder to have the same candid discussions or even ask him for support or guidance. Plus, more factors started to come into play, such as **toxic and divisive behavior, which I'd hoped had died off** with the last project. Trends and behaviors like what I experienced at Champion Mechanical started to show their ugly heads.

James had a best on-site buddy—the interior superintendent, Mac. **They went way back to jobs they had together at their** previous company. James always touted how he got Mac a job with our current contractor, like it was a badge of honor. We all shared a room in the trailer. It was a decent-sized space for all four of our desks, which were lined up against the walls. Breck and I sat right next to one another.

James and Mac cracked jokes all the time, making fun of others, shooting the shit about other projects they were on, and even badmouthing the executives we worked for. James would make loud comments about how he didn't care who was listening, which always indicated the opposite. Their jokes went from inappropriate middle-school to full-blown and overly descriptive of women's bodies and attitudes.

I was increasingly uncomfortable while sitting at my desk most days. Forced to listen to them complain about their wives or comment on the good or bad skills of other women we worked with, I started to feel extremely self-conscious.

This was my first project working with so many women in leadership, from our subs to our contractor team. James complained about the "poor ability" of our only female project manager, who managed our structural scopes. He'd pull up rap music and blast it on the TV behind his desk while we worked. He'd sit there and refer to people on our trade partner teams as idiots. The list goes on and on when it comes to the negativity he brought to the trailer.

I noticed a shift in my behavior, including wearing looser clothing, keeping my distance from both the project manager and the superintendent, and asking for help only if I absolutely needed it.

The internal turmoil I felt from the anxiety and mental battle created self-doubt as well. Their commentary wasn't directed at me while I was in the room, but if they spoke of other women with such disregard, what kept them from doing so about me when I wasn't around? After a few months of dealing with their unprofessional and harmful behavior, I finally called the anonymous HR tip line to report James.

Site Photo with Portland Team

About a week later, James ended up having to do a special harassment and ethics training. I knew this because he sat there and complained about it loudly. He kept harping on how sensitive people were and how no one could take a joke anymore. So now, he said, he had to take "some dumb online course." His temper and demeanor tampered down a bit, but only for a week or so. Then he was right back at it. Nothing changed, and I was still stuck in that crummy spot with a male manager I couldn't talk to. He was the one person I should have been able to turn to. Instead, I had to take whatever he dished out.

My relationship with James and Mac continued to fray. Meanwhile, added frustrations were going around due to project delays, poor communication, and misalignment with the architect. I started to get scolded more frequently. I spent a significant amount of time on my work but was told I was doing a shoddy job.

This was, of course, in conflict with my relationships with and the feedback I was getting from the trade partners I worked closely with. They would comment on how detailed I was and were attentive to conversations. I had built strong relationships and partnerships with them that James was not able to build himself. My approach was fused with listening, engaging, and asking questions. More importantly, I was willing to roll my sleeves up and problem-solve together.

For Women in Construction Week, one of the subs I worked with gifted me some swag, including a nice silver Nike zip-up, which I still have in my closet today. For Christmas, they gave me a nice bottle of wine and a sweet holiday card. This was all given to me in front of James, which made him angrier and widened the gap in our partnership.

But there were some significant bright spots. I befriended a few of our other field engineers and project engineers. Adrienne was a delight, with her boundless energy and optimism. She and I had great conversations as we walked the site, learning about each other's scopes and sympathizing or problem-solving together during rough patches. She even had me and a few others over to her apartment for Halloween and game night. She was just as passionate as I was about making the job site and the company more inclusive for women and other minority groups.

Over time, we formed comradery with a handful of other Momentum Builders women who were working toward a more robust **Affinity group, similar to a business resource group for** folks who share common interests or professional challenges. Together, we were proud of the events we put on and the resources we continued to implement to connect the PDX group in new ways. Hosting lunch and learns and bringing in other woman leaders to talk to the group of us was inspiring and made me feel hopeful. I found pride in the DE&I efforts I was supporting, and that filled my cup in new ways.

STRUCTURAL STEEL

The Portland office hosted a monthly field engineer/project engineer site visit to create visibility and exposure to one another **and the work being performed on different projects.** That little bubble grew bigger with each encounter, building a network. I took this opportunity and even engaged with other leaders and project managers, visiting their project sites and meeting with our director of operations for lunch. Outside the project I was working on, the company felt inclusive and caring. I continued **to find ways to be energized and supported to grow in my skills** and management capabilities.

This was great for a while, and it fueled and somewhat distracted me from the growing difficulties I was experiencing with James.

One day, I got pulled into our large conference room in the trailer where we hosted many of our architect, engineering, and contractor (AEC)[22] meetings or project schedule updates. In the room sat my current project manager, James, and our structural steel and concrete project manager, Casey. Immediately, I had goosebumps and mentally prepared myself to get fired. Instead, I was stunned by an invitation to lead up our structural steel scope.

The previous project engineer had taken another job with a different contractor. They thought that, with my background as a welder and my experience working with a mechanical contractor, I'd be a good fit. Initially, I was elated to get away from James. I looked up to Casey, who was the only female project manager on our team, as a role model but couldn't find a good way to connect with her.

I took the opportunity and did my best to turn over to Breck the scope I was managing. This included contract documents I had built binders for, RFIs that had been submitted or saved, and other approvals that were required. This new opportunity felt like the golden ticket to get away from James and move into a more prestigious role, make a bigger difference, and be seen for my skills.

But shortly after moving over to the structural steel team—one of the major scopes Casey was leading—I quickly realized the mess I'd walked into. I liked the space I was provided to work with, which included two desks or long working tables

---

22 Architect, engineering, and contractor (AEC) meetings are half-day sessions that bring together key decision-makers to review the project process, ask questions, and resolve problems.

dedicated to my drawings, binders, and other tools needed for the job. I even had my very own whiteboard to map out ideas, priorities, and to-do items.

The painful week of transition quickly set the tone for what I had been brought into. The project engineer I was replacing, Meegan, gave me few to no details. Whenever she and I walked the site, she skipped around questions and even bounced from one item or element while we were doing quality inspections. Her idea of an inspection was to count how many steel precast or beam anchors were in place. She did this with no print in hand or an iPad with Procore open.

After about day two, I observed two things. One, she reeked of alcohol the whole time she was on-site. And two, she didn't give a shit. She would show up in cute wedge boots and low-cut blouses, with no intention of stepping foot on-site. I was the opposite, wearing blue jeans, Redwing work boots, and a polo.

As someone who came from the trades—touching the work, talking to those in charge of making the scope a reality, and observing how the construction was being completed—it all helped me get a fuller picture of the responsibilities I was charged with. Meegan made no effort to coordinate with the active subs she was supposed to be working with. Turnover documents didn't exist, and she deleted every single email so there was no work or correspondence history. Her approach felt irresponsible and careless.

Over time, I came to understand why she was acting the way she did, but it took a few months of working with my new team to gain those insights. Time went on, and I quickly became associated with all the challenges intertwined with the scope I was managing and even the subcontractor we were using. I did my best to navigate those relationships by being present with the crew during stretch-and-bend, the dedicated start of the morning where all crews come together to listen to important safety updates and literally stretch for the workday. This was always one of my priorities before heading over to the trailer—checking in to bring new info or updates. On some occasions, I even brought donuts.

All of this was not enough. Things rapidly unfolded as the work became messier and more chaotic. One item after another became problematic. Our embeds (fancy metal bases used to weld metal beams or structural elements onto) were increasingly uncovered in the wrong spots due to incorrect dimensions or over-settling of concrete.

Remember the earlier point I made about the project engineer, Meegan, I'd replaced? She didn't coordinate well with the concrete crews or perform quality inspections.

The majority of this problem came from a mix of the structural team and the associated subs not working well together. There was lots of misinformation, many toxic exchanges, and little understanding about the drawings designed by the architects and engineers who stamped them. All the work I took over became my fault for not being done since I was the face now associated with the steel scopes.

The superintendent I was working with at the time was an old curmudgeon who'd been in the field for more than twenty years. I would ask safety questions or inquire about certain documents we should use to set the teams up for success, and he would dismiss me or say, "No, I don't need anything like that to make our trades work well."

A few weeks later, I watched in horror as a boom lift being craned up to a higher level resulted in a safety incident. The boom lift was not properly hooked up with straps or locked in place, which caused it to flip sideways as it was lifted. Had the stubborn superintendent done the requested work and used the correct planning methods to ensure the steel structure scope was well coordinated, this wouldn't have happened.

Soon after the incident, the same superintendent told me I needed to start coming to the site at five a.m. to do stretch-and-bend with the first crew. Emotionally stressed and edging closer each day to burnout, I pushed back a bit. But he yelled at me for arguing. I went home that night bawling like a baby, utterly exhausted but expected to do even more.

I attempted to recover some of the lost progress in working with the project manager and let Casey know my concerns. Most days, though, I was running a hundred miles per hour, chasing down RFIs,[23] sending over marked-up Procore[24]

---

23 An RFI (Request for Information) is a typical communication tool between contractors, engineers, or architects used to ask a question or potentially suggest a change to a scope of work.

24 Procore is industry-standard construction management software. It's used for scheduling, reviewing drawings, and communicating changes to a project. Request a demo. Procore. (n.d.). https://www.procore.com.

drawings to work with the engineer on record, or being the punching bag for the subcontractor. Every time I stepped on-site, I felt increasingly insecure about the work I was managing and the team I was supposed to be working with. Much of the time, I was being yelled at or spit at. At times, I even had tools thrown at me because I was one of "too many cooks in the kitchen."

Over time, I made some strides in building a relationship with the only female ironworker on-site. She and I bonded over our love of women. It took some time, but both of us figured out we were both part of the LGBTQIA+ community. She shared stories about what she did with her wife and children over the weekend. Sometimes, we talked about our first girlfriends. We both appreciated the safe space and commonality.

But even she started to create distance from me as well. Soon, our comradery quickly dissolved. It didn't matter how often I went to bat for her or argued with the general site foreman to get a female-dedicated porta-john to the top level of the project. One of the foremen had initially taken a shine to me, even to the point of calling me his work wife. (Yup, it's gross, but it was still a win in my book.) Even he started to distance himself from me or would constantly complain about other trades or delays.

Anytime I tried to do an inspection, I was scoffed or cursed at. It got so bad that I tried to form alliances with other field engineers and project engineers from different trades to walk the project with and collaborate on scopes. I was feeling more and more helpless with the ship that was sinking and delaying the beginning of other parts of the project.

Major material delays became another big part of the structural steel scope pain points. The project owner (client) we were doing the new build for wanted a very specific set of blue steel[25] stairs that would connect two levels and be the focal point inside the building. This was an expensive endeavor to begin with, since everyone from the architects to the customers had to fly down to Tennessee to investigate blue steel options. That same steel had to be cut, welded, and then shipped out to Oregon.

This long-lead-time item became a bigger schedule delay than expected, and it snowballed many other scopes. The increasing stress and anxiety on site were palpable. Safety incidents went up, and my crew gave up any attempts to wear personal protective equipment (PPE) correctly. Sometimes, they didn't wear it at all. Not wearing proper safety gear meant more folks were getting hurt, which fed into the negative perception of the project. Many people were grumbling as they talked about the slipping schedules and increased badmouthing of other trades. In many ways, teams in the field were shrugging, as if they felt that no matter what they did, it wasn't enough.

---

25 Blue steel, a specific mix of alloys and external processing, is typically very expensive due to the special treating of steel. *Blueing of Steel*. Custom Machined, Forged, Cast & Plated Parts - Bunty LLC. (2020b, May 29). https://buntyllc.com/blueing/

They seemed to not care about being safe. Others claimed the safety gear was cumbersome and slowed them down. Well, so does a broken back when you fall ten feet. And so does partial blindness when you lose an eyeball because you didn't want to wear your safety glasses.

This did not sit well with me since I took safety very seriously. As a welder, I had hurt myself plenty of times—including in high school, when I was learning to stick weld and burned a hole in the left sleeve of my black hoody, which left a gnarly scar. As a welder working in a shop building bus frames and out in a garage fixing up hot rods, I had screwed up my back lifting jigs that were too heavy. And the repetitive motion of pulling a trigger for eight to ten hours at a time had ended up causing me to develop carpel tunnel syndrome.

So, when I walked the site, I took safety seriously. It was our company's biggest message: everyone owns safety. I had to tap my safety glasses at folks too many times because they had their own hanging under their chin or sitting on top of their head. Or they weren't wearing them at all. To help mitigate this behavior, working with the on-site safety engineer, I carried spare glasses, gloves, and earplugs. Eventually, we got a PPE vending machine housed in a midsize blue sea crate to help create better access to required safety equipment.

However, this wasn't appreciated for two reasons. First, I was docked points on my annual review, which James (my former project manager) and Casey (my current structural steel project manager) held. James took the lead on this evaluation and told me I was too direct in how I communicated safety protocols on-site. Casey elaborated, saying I needed to be nicer when I talked to folks in the field. Her example was that I should ask how someone was doing or what they were working on, then bring up their lack of safety glasses.

Interestingly enough, James had brought up this same point regarding my emails. He said my messages were too direct and needed to be nicer and less firm.

Many of the responses I received from the structural steel trade partner I was managing fed negatively into my annual review. Regardless of the two or three men who were consistent in saying, "Fuck you," and "You don't know what you are talking about" whenever I'd try to redirect a potential safety concern.

One man in particular, who was angry at my insistence on safety gear, walked off-site. His reasoning was that the Hawaiian shirt he was wearing was an attempt to brighten up the bad environment. I told him I was all for the shirt but that he needed his vest. This didn't fly for him and became a whole argument.

My point to him was that keeping himself safe meant keeping other trades around him safe as well. I kept my cool and stood my ground, but he still stormed down the ramp and over to the trailer, then proceeded to leave.

Unfortunately, this foreshadowed a much darker road ahead.

Within a few days, this same ironworker, who had put a target on my back, started to follow me around the site, smirking or making snide comments. I would be in a middle of a conversation reviewing plans or setting up site photos, and he would come around the corner, bump into me, and call me a "dumb bitch."

One day, I was walking with our integrated construction coordinator, Adeline (a.k.a. visual guru), who could show Revit modeling on how the project should look once built or use software technology to identify potential conflicts with the design. We were walking the site and looking at a few things she wanted to inspect when we passed one of the crews I supported. They were setting up their harnesses so they could hang down through a floor opening to weld up some steel. I casually went by and asked how things were going with some hope of repairing this frayed partnership. I asked a few questions about the safety gear, including where folks' vests and other items were. This was met with a tossing of tools and talk about me in the third person. I nodded at them and replied that I understood, then continued the walk with Adeline.

As we were headed down to the lower level, still within earshot of the crew, the ironworker who had put a target on my back said, "Man, she just wishes she could do my job."

Adeline turned to look at me and said, "*You* could do their job, couldn't you"?

I slowly nodded and said, "Yeah, and at times, I did. Just in a controlled environment, not on-site."

She gave a partial smile with a somewhat proud but sad look in her eye. We both recognized how, in one sentence spewed by a man, we'd been dismissed as women but also as managers.

Things continued to get worse, with said ironworker bumping into me and muttering loud, crass[26] comments. My anxiety continued to increase, and I started to feel concerned for my well-being. When I was working as a welder, I had heard horror stories about women having tanks dropped on them or tools being tampered with. The evolving hostility from the crew I was supposed to be supporting was hitting a brim, and I had no way to fix it. I started walking the site with others and would ask different team members to go walk or inspect something with me.

Eventually, word got out to Casey. She wanted to know what was going on and why. When I told her about what was happening in confidence, mentioning that I didn't want escalation and that it would only come back and make my situation worse (*which it did*), she said we could handle it quietly.

Casey requested that I send over a statement explaining the situation. So I sent it over. She immediately said it was too

---

26 Men will verbally assault or write profane comments, calling women names, such as bitches, hos, etc., on a job site. These words can also be seen written in Sharpie on temporary wall framing or with a finger on a dirty window. In the porta-johns, I would see pictures galore of male anatomy. On several occasions, we had reports of someone smearing human feces all over the seat, door, and whatever else they could touch inside a porta-john. This is why the women's porta-johns are often locked. It gives women the hope of having a dedicated space to take care of business.

wordy and that I needed to shorten it. After going back and forth a handful of times, I finally pared it down to one sentence: "[Name of Ironworker] is harassing me on-site."

This is what she sent over to the trade partner, our legal team, and HR. A week later, the ironworker in question was removed from the site. No repercussions or punishment. He was just moved to another project. No dock in pay or loss in time. Just relocated. Simple as that, with no accountability.

Within the same week, I started to get stepped on during stretch-and-bend. My crew was intentionally forming a wall of bodies so that either there was no space for me in the large group or men were standing so close to me that I couldn't move. When I would move to stand with the crew, different men would sometimes walk over to the opposite side of the level where we met every morning.

It was clear that this arrangement wasn't working and would only get worse. I couldn't carry on like this. Even after having heartfelt one-on-ones with Casey and talking with others in confidence about my struggles, I saw no end in sight. I had to act.

Things in my personal life were also cracking, with a romantic relationship in shambles and my mother's health fluctuating out in Minnesota. So I laid the groundwork to transfer back to the Momentum Builders headquarters in Minnesota.

After sitting down with our office's director of operations to highlight some of my professional accomplishments and personal challenges, I was able to get relocated back to Minnesota.

OREGON

By August 2019, I was back in Minneapolis and had been placed on a six-floor commercial tenant renovation downtown. I was going to lead a previously occupied space in the downtown AT&T Tower that was being repurposed for a new client.

CHAPTER 7

PANDEMIC

What is happening?
The world is shutting down.
How do we navigate the unknown?
– One day at a time.

𝒰pon my arrival in Minneapolis in the late summer of 2019, I was giddy for a change and eager to get back to a somewhat familiar place. And I was happy to be close to my mom again. I was looking forward to the change of pace— and to not need to break up site fights between trade workers, collect piss bottles to throw away, or hear about the gallon jugs of pee found in the crane. I was ready to return to a familiar place—back in Minnesota—and have a new team. Trying to support two homes had been stressful and, at times, hard to manage. Optimistic that this was a new beginning that would allow me to finally grow my skills and embark on more inclusive challenges, I was ready to begin anew.

Being tasked with a smaller project worth a few million dollars, rather than multimillions to almost a billion, gave me new opportunities for greater responsibility. Construction projects are based on the estimated price to build, i.e., $2 million vs. $22 million. The increased price rapidly changes the length of the schedule, the overall scopes that need to be managed, and the size of the team to support. Working on a smaller project meant I would have more autonomy over key decisions and work much closer with executive leadership.

From the start, I was offered full control over the project along with a construction manager named Lexi. Both my new director of operations and project executive encouraged Lexi and me to own the six-level renovation downtown. This was a delight to hear, and I finally had a sense of independence that matched my title and experience.

Together, Lexi and I were going to be managing a tenant renovation project in downtown Minneapolis in the AT&T Tower. The only thing I knew in advance was that this was supposed to be a short project of under a year that would have a total of three folks managing it: Lexi and me as the construction managers, and Owen, who was the site superintendent.

The first week of the project seemed like a breath of fresh air. I was able to work closely with Lexi and Owen. The three of us seemed like a capable bunch to run a midsize project. Our scopes were smaller but pricier, with some significant architecture and engineering changes to three of the six levels we were managing. There was a design implementation to install heavy metal stairs to connect an open floor plan between the main level all the way to floor three of the commercial tenant office space we were renovating. The floor plan throughout the levels was relatively simple, with much of the work being focused on tearing out or removing the previous tenant's remaining space.

Old graffitied walls and bathrooms on every floor that needed to be reframed were just a few indications of the renovation work ahead. The expensive parts were the Italian tile they wanted in the bathrooms; the special wallpaper, which we had to rehang multiple times; and the special kitchenettes.

I quickly developed a strong relationship with most of the crews. My attitude when it came to safety was similar to how I'd approached it in the past. But in the Midwest, it was much less of an issue. People didn't scoff or immediately get butt hurt because I asked them to put on their safety glasses. It was a nice change of pace to feel more at ease with my job and some of the scopes I was managing. This was the first project on which I was brought in closer to the beginning so that I was able to see it through to the end. The project carried on with consistent safety calls, a two-week look ahead, and architect, engineer, and contractor meetings.

Things were flowing well for the first month or so. I was mostly on the job site all day, wearing my safety vest with my tools, tape, and iPad serving as a mobile office. Again, I practiced what I'd done in California, which was being present where the work was being done, asking questions, and helping resolve challenges.

At some point, things started to shift. The Momentum Builders group I was working with had its carpenters come on-site. A large group of them were crass, engaging in visually unappreciated actions, like grabbing their junk at stretch-and-bend, loudly cursing each other out, and even commenting on **female bodies. One particular foreman was incredibly defiant** with his actions. He would talk back when someone asked a question or provided direction. He'd roll his eyes and even **gesture at Owen, who would laugh it off.** This quickly wreaked havoc on the vibe of the job site, making it more hostile not only for me but for other women and some of the trades.

Additionally, we started to slip on our schedule, partially due to long-lead-time materials and having to redo mockups for the architect. I also started to recognize a trend when it came to communicating information and project status. Lexi, who I'd been fond of initially, started to get snippy with the way she **posed questions or offered information. A few times, she even** set me up to fail during meetings.

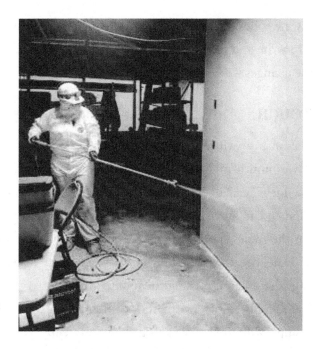

Jessi painting with sprayer for the first time

For example, When I came back from one long weekend, we had a morning check-in meeting with our project executive. She mentioned an agenda item that had to do with my scope but provided no relevant details or updates beforehand.

We had the meeting to run through the updates, but instead of owning the information she knew to be true (which had changed while I was out), she immediately put it on me, saying, "Jessi will update us on this item."

I felt like a deer in headlights, and the only thing I could say was, "I'm not familiar with those changes and would need to coordinate with the trades after the meeting."

This gave Lexi the spotlight to shine. She stood up and started to write and draw on the whiteboard, giving specifics of what the changes were.

BAD TOUCH

My relationship with Owen also became strained. Close to my two-month mark at the company, we got an assistant superintendent called Red, who was much calmer in his demeanor and in how he handled challenges on the project. Many of the trades started to come to him and me when they had an issue because they knew we would help instead of throwing a hard hat or cussing them out.

Around this time, with the combination of strained relationships with my contracting team, the slippage in the schedule, and increased trade activity, the wheels quickly fell off the bus. Owen, who was a twenty-plus-year construction veteran, started to get more aggressive at our two-week look ahead and our morning stretch-and-bend.

One day, I started to walk over to the mouthy carpenter foreman to point out his lack of engagement in the morning stretch-and-bend and his missing PPE. He was setting a bad example, and I needed to say something. I mentioned it quickly to Owen in passing, and he said, "No, don't."

As I walked, determined, toward the foreman, I felt everything change in an instant. My heart started racing, and my face flushed with heat. Then it happened. In those brief seconds after his comment, Owen grabbed my right forearm firmly, jerking it abruptly in an attempt to stop me. I glanced at him, stunned that he had touched me in this way, let alone so aggressively. Shaking off his grasp, I continued to make my way over to the foreman.

This all occurred in a matter of seconds. Shock waves went through my body. I was astonished and internally grappling with the physical action, wondering why he would grab me in any way, and especially in such a manner. I froze in my tracks and decided to go busy myself elsewhere to create distance and reflect. This incident had been witnessed by almost everyone standing there after stretch-and-bend, including our assistant site super, Red, and all the trades.

After mulling it over, I knew I was right in doing my job to protect every crew member on the project. Owen had no right to put his hands on me. Later, while I was walking the site, a few other trade partners commented on the interaction and asked if I was okay.

Internally, I was not. But of course, externally, I kept a calm, happy face and said all was good.

Indeed, it was not good. In a matter of two months, I had gone from feeling like I was making progress in my career to being completely mortified and shell-shocked at the turn of events.

The next morning, I decided to nip the incident head-on by addressing Owen to his face, and I timed it so the assistant superintendent, Red, was present. If I was going to make any sort of impact with my stance to set boundaries—while also not leaving the interaction to be characterized as a she-said, he-said situation—I needed a witness. Plus, Red and I had a growing on-site partnership. I looked to him as a mentor in many ways and assumed he would have my back if things went south.

First thing in the morning, once I knew all present parties **were in the same location in the site office, I walked into the** assistant and site superintendent's joined workspace. Standing directly in front of Owen's desk, squarely planted and making direct eye contact, I said, "Don't you ever put your hands on me again. You had no right to physically touch or stop me from doing my job."

As one might imagine, this did not go well. Owen stormed over to me, got inches from my face, and looked down at me. I **didn't flinch. Instead, I said, "Don't touch me again, and don't** stop me from doing my job."

He went on about how I didn't know anything and how it wasn't my job to speak to the crews about such matters. Our project executive, Caleb, ended up involved. Owen went to him, either insecure about how I would respond or wanting to elevate the incident. When I met with Caleb, he asked about what happened and how I would like to proceed. My response was simple. I had addressed it verbally with witnesses and was **confident the matter was handled.**

He recognized my wish to not escalate and told me that, for now, this was where we would leave it. If other issues were to surface, I was told to let him know, and he would take it to HR.

PANDEMIC

The situation continued to get worse from this point on. Owen continued to find moments when he could corner me while no one was around. At one point, I was filling out safety reports at my desk, a customary contractor expectation and a practice I'd performed repeatedly in Portland. Owen stood right behind me. I could feel his breath on my neck and see his reflection on my computer screen.

He asked, "What do you think you are doing?"

I turned to look at him, making sure to make eye contact, and said, "Filling out my safety report for the week."

He asked, "Why?"

"Because it's my job to do so."

I turned around and continued to fill out my documentation on our internal safety site. In my mind, this was a strange thing for him to question, especially since, company-wide, the message was that everyone owned safety.

Owen remained close to my back and desk chair, then stated, "No, that's my job." His tone was low and growly, and he towered behind where I was still seated. With arms crossed and a scowl on his face, his body language was threatening. If you have ever been in an abusive relationship, you know what that looks and feels like. It's as if at any moment, they might take a swing at you or cause physical harm.

I explained that this was what I had been trained to do in Portland and that it was a company-wide expectation.

He continued in his low tone. "No, that's my job. You are no longer permitted to submit those without my approval."

I politely responded, "I do not require your permission to do my job or report safety concerns from the week."

At this point, Owen seemed to be further irritated at my responses and refusal to abide by his authority.

I gathered my site gear and excused myself. Squeezing out of the tiny office without touching or brushing him, I climbed to the highest level of the six-floor renovation we were working on in the tower. I used the staircase to travel up the floors instead of the elevator, knowing it would be close to dead, with no one around that late in the afternoon and most crews done for the day.

I stood there shaking in silence.

PROBLEM CHILD

The ongoing escalation with Owen started to interfere with the work being performed on-site. There was a major milestone project coming up to install three levels of stairs connecting the core of the tenant space for the client. About a month in advance, I had recognized this work as a hazardous scope, not to mention the passenger-sized elevator we had access to for moving material in the tower. I brought this to Owen's attention. He was the site superintendent, so it was his job to help protect and **manage the daily activities of the fieldwork.** I wholeheartedly thought he would want to plan this early.

I was wrong.

He argued with me, saying there was no need to plan it out. He said we would be loading everything up early in the day, before the trades showed up, and we'd be done well before the elevator could get jammed up.

I was not confident in this suggestion and slowly worked with a few of the trades on the back end.

First and foremost, I made sure to consistently remind all scopes of the day and time the steel would be coming in. I even worked with the trades themselves, asking questions about how much material and weight would be involved, then confirming and reconfirming the date of the delivery. The steel ended up delayed another two weeks but managed to finally arrive.

So many things snowballed on that late fall morning when it got there. First, the trade steel sub, showed up two hours after they promised the delivery. And instead of coming through the back, they showed up at the front door of the tower, where employees, business owners, and customers came through. The passenger elevator quickly got backed up because tradespeople had trickled in to start their day per usual. We ended up with steel all over the tower lobby floor. Steel beams were bunched together and staged as neatly as possible, then covered with tarps. To say the owner of the building was furious is an understatement. It was a mess that took twelve hours to fix.

All the work I'd tried my darnedest to accomplish failed. Simple coordination and support from Owen might have helped us eliminate the added frustration and mishaps. This was a direct reflection of how disorganized we were. In all honesty, the steel should have been sent back instead of offloaded, or there should have been a special agreement to stage it in the loading dock instead of spreading it over the entire tower lobby for all to see.

Things went from bad to worse. In a matter of a week, Owen went from a mix of aggressive to downright hostile. Late one afternoon, when I was walking the site after finishing up a call with a lovely female superintendent in Seattle, Owen walked up to me and got in my face. With his chest touching mine, he stood over me like I was a small child being punished for touching the stove while it was hot.

He was irate for some reason and started to call me out about my attitude and about how I never listened to anything he directed me to do. He was literally breathing on my face, with his eyes popping through his safety glasses.

I didn't move an inch. I let him spew off the streams of semi-coherent statements about my inadequacies. Finally, in a calm, lowered voice, I responded to his accusations.

"You are pissed off at me because I had a fucking career before this one. I came onto this project well-versed and knowledgeable about safety protocols, and I actually know how to do my job well. Your anger stems from the fact I have a better relationship with our trade partners than you do."

Finished but also aware that he might strike me, I quickly turned and headed down to the lower level, where people were actively working. After performing a few inspections, I finally went to the bathroom on a public level and hid in a stall, crying.

It had happened. The final crack had occurred, and I could no longer handle being surrounded by his threats. After taking a significant amount of time in the bathroom to collect myself, I made a very hard and courageous, yet impactful, decision.

I needed to report Owen for his harm.

I quickly tracked down the assistant superintendent, Red, whom I'd had a pretty stellar relationship with until about a month or so prior. His body language, conversation, and overall engagement with me had gone from enthusiastic to distant. He'd stopped making eye contact and having bigger discussions about topics. It had been so fun to talk to him about different aspects of the project and my future aspirations of becoming a site superintendent.

Not having anyone else to turn to, I called him as a last-ditch effort. To my dismay, I found he had already left the job site for the day. Shaking, I told him about the interaction and said I felt threatened and afraid to be on the site alone. Even the other project engineer was gone for the day.

Clearly annoyed and distracted by some family challenges, he told me to hang tight. About ten minutes later, I got a phone call from our project executive, Caleb. He asked me about the situation and told me to stay put a little bit longer, promising to call me right back.

As I was quietly hanging out in an unfinished bathroom on a higher level where I was sure to not be discovered, he called me back. He told me Owen was aware that I was leaving for the day and that he would handle closing up the job site. He continued to instruct me to head home, collect my thoughts, and come straight to HQ first thing in the morning. Caleb also informed me I would be meeting with our director of operations, Kaleigh.

Initially, this gave me some sense of hope, since I considered Kaleigh a mentor of sorts. She had brought me into planning for the Women in Construction events, sat with me to review my board of directors (tool for sponsorship) which is made of leaders or other industry experts to help me take my career to the next level. Kaleigh seemed to know where I was coming from.

I left quickly and went straight home. I could feel only a mix of numbness and disbelief. I started to question the abilities, **confidence, and knowledge I had worked so hard to build.** By default, I was blaming myself for the abuse Owen was dishing out. He was punishing me for doing a good job. His insecurities were surfacing as the project schedule slipped, budgets became askew, and trades stopped listening to him. I guess he had thrown one too many hard hats at our two-week look ahead. **Folks were done and gunning to get off this small renovation tenant build.**

The next morning, with my notes in hand, I met with Kaleigh. At first glance, we had a pretty good relationship. On occasion, I would meet with her to engage and learn and gather as much mentorship as she was willing to offer in my new work setting.

Since this was an escalation from the previous incident with Owen, we had to document what happened. She asked a lot of questions, trying to pinpoint what exactly Owen was doing and how it contrasted with how he'd been in the beginning. She wanted to know specifically what it was he was doing that was making me terrified to be alone on the site.

What initially felt like being supported and sharing a commonality of experience quickly morphed into being questioned about my part in the harassment. Somehow, it was my fault. I'd caused Owen to distrust me to the point he had to micromanage every action or task I completed.

I was feeling very unsure about this conversation that I had come prepared for. Kaleigh appeared to be intently listening, but her body was closed off. Her arms were folded in front of her, and she was leaning back in her chair. She'd gone from asking, "What's going on?" to "What did you do in this instance? Why did you push back on his authority? Was it a good thing to proceed when Owen told you not to?"

Toward the end of the discussion, I felt gaslighted and was questioning the events leading to that morning of the conflict.

This triggered an HR investigation, which meant Owen received explicit instructions to not speak to me or engage whatsoever.

Lexi and Red decided to take the same stance. The victim became isolated and further removed from the details of the project. It felt horrible at first, with awkward silences and sideglances that weren't accompanied by actual expressions or words.

Then I started to relish the quiet. For the first time since I'd started on that tower project, I was free to coordinate, collaborate, and inquire without a project engineer or superintendent asking questions or verbally pushing me aside to take over a task or undercut the point I was making during a meeting.

DÉJÀ VU

A month later, I was put on a performance improvement plan (PIP). Instantly, I had a flashback to California, when Eli and Bowen sat me down to talk about my inadequacies on the refinery project. The memory hit me hard, almost like déjà vu, when I was sitting with my project executive, Caleb, as he informed me that my work performance was suffering and that corrective measures required regular check-ins with him and Kaleigh.

I was stunned. Recovering as best I could at the moment, I asked for specifics. They made mention of meetings where the other project engineer, Lexi, had to "rescue" me and complaints from folks and my contracting team, who said I spent too much time on-site instead of in the office.

The final bit that he drilled in had to do with the off-site meetings I had with other colleagues to learn about their projects. He said I was a great networker but needed to focus more on the job at hand.

It felt like the same repeated cycle I had experienced over a year ago back in San Francisco with Champion Mechanical Contracting. I was not fitting nicely in the box they wanted me in. Ambitiously, I was doing all the right things to learn and grow, but somehow, I was being punished for it.

PANDEMIC

There was some foreshadowing to this event that I'd missed early on.

One early morning before the crews showed up, around five thirty or six a.m., I was walking the site, scanning my emails, and collecting project updates to share. I was headed out to the site when Owen was standing in the main elevator lobby, which was the public shared space on our level. I didn't catch much, but at the end, I heard him say, "Thank you so much for your support. She's just a problem, and I don't know where she makes these accusations up."

Doubling back to the office, I took a moment to reflect and quickly put two and two together. Owen had rallied others to support him during the HR investigation. He had been with the same company for almost twenty years. I was just some twenty-year-old with no network or strong sponsors to go to bat for me. In essence, I was screwed.

The PIP continued with me coming to every check-in prepared. For the initial sit-down, I came in like an attorney prepped for trial. I looked nice, with jeans, a black blouse, and a festive green velvet blazer. My portfolio contained references to incidents, details about project successes, and even emails between myself, the superintendent, and the project engineer. The idea was to paint a fuller picture of the environment I worked in and to show that I was more than willing to do a good job but was constantly being barricaded or limited via communication or lack of information.

It was a fine line to balance, not coming off too defensive or making excuses. Being prepared helped me feel less anxious about sitting in front of the judgment committee. (Kaleigh and Caleb were determining my career, after all.) This went on, back and forth, for a few months until something almost miraculous happened.

At the end of 2019 and the start of 2020, COVID-19 began to surface. People had no idea how to treat it or what to do. Quickly, the focus went from me to what to do now. Confusion ensued, and uncertainty ramped up. By this point, the project was finally getting close to completion. Trades were moved off-site, furniture was being staged, and finishers were working with me to complete punch list items.

The end was in sight.

I had a scheduled sit-down with Caleb and Kaleigh, which ended up being my last. It started with Caleb taking the lead, making comments about my performance and how he appreciated my openness to improve. About ten minutes in, Kaleigh removed herself, leaving Caleb and me alone. He rattled on a bit longer about how I was steadily becoming more of a team player, etc. I took the final bit of the sit-down to be courageous and put into perspective all the work I had accomplished and the process I had followed.

Knowing he wanted me to bring lessons learned, I detailed, with the document in hand, each thing that I'd actively done, but I reframed them as the things a strong construction manager does on a job site.

At the end, he was happily nodding at the wealth of knowledge I had gained from this PIP. Somewhat shakily, I firmly said, "These have been and continue to be best practices I have been implementing since my time at Kinesis Contracting began until now on this project at the Tower."

With some mumbled sentences and slight disagreement, he struggled to have a significant response. I made sure to thank him for his time and went on with the rest of the day.

Two weeks later, I was told I was getting moved from the project and was starting my scheduling rotation at the beginning of April. This was a small victory in my book. Our assistant superintendent had already been moved off the project, and most of the punch list items I had walked were primarily done. The last big push consisted of assembling and installing the office furniture and decor.

My last day was an internal celebration as I happily cleaned up my desk, loaded up the car, and handed over last-minute documents. At the morning stretch-and-bend with the much smaller remaining team of scopes, I said my goodbyes from our distanced stances. I shared how much I had learned and grown from each of the trades and their patience as we neared the finish line. After completing my final site walk and tidying up final questions or tasks, I started to make my way out the door.

Lexi didn't seem super thrilled with the glee I exuded over being done with our renovation project, and she was clearly trying to stymie it. As I was putting my jacket on, ready to head out around lunch to go home and set up my home office, she burst into our shared space and told me I needed to be on a mandatory site walk with the designer.

A bit confused and caught slightly off guard over a scheduled site walk on a Friday, I said, "I was heading out." I didn't see any reason she couldn't handle the walk by herself.

Her face turned the color of a beet. She emphatically told me I didn't have a choice in the matter and that I needed to be on this walk.

Still struggling to understand why, I said, "Everything is in **Procore. Why do you need me to walk with you to look at final punch items?"**

She replied, "Caleb wants you to be there."

Still lost, I looked around and didn't see him. So I said, "All right, let's go. But I need to leave soon."

When we met with the designer, he immediately asked a few questions directed at Lexi. Looking rather discombobulated, she turned and stared at me, waiting for my answer.

Having worked intimately with all our trades over the approximately six-month project, I started to answer a few of his questions, then paused. My iPad froze up, making it hard to **look up specifics.**

Lexi had nothing more than a work phone in tow, but I asked her to look up a few things and again watched her face **flush red. I quickly realized why she wanted me on this walk.** She didn't have a clue about what was going on or where we were in the punch list of closeout items, and she needed me to soften the appearance of her cluelessness.

Relishing her self-imposed hardship, I tucked my iPad under my arm and said, "I am working blind. Lexi, you'll have to take it from here."

From that moment on, I stood back and watched her revisit items the designer pulled up on his tablet. As he asked questions and she fumbled responses, I kept my face as emotionless as possible. In one fell swoop, I realized how much I had truly accomplished on this project.

All those instances when Lexi had tried to make me look bad **weren't because I was failing to do my job. They were because** *she wasn't doing hers.* Any opportunity she had to not communicate or prep before a meeting was so she could reposition herself as knowledgeable. Realistically, she was setting her team up to fail.

When I drove away from downtown Minneapolis that day, I felt victorious, even for a brief moment.

I wasn't the problem. They were.

Over the next two months, I focused on my scheduling **rotation. This was a new experience in many ways.** Learning to work remotely and trying to not feel isolated, combined with having to develop a new understanding of how to build a project schedule, was challenging. I was grateful to have my mom living with me and my fur baby Suki close. Both eased the anxiety I felt daily.

The process of learning scheduling remotely was frustrating for a few reasons. I was working with P6 software, which is a complex system of connecting scopes of works through a Gantt chart-like approach. It looks like a complex series of lines that are busily connected on a screen, all used to schedule various scopes and milestone work.[27] It was quickly confusing for someone who had limited scheduling or software experience. In theory, I understood the natural flow of work, but when it came to planning, I preferred and better understood two-week look ahead and Takt scheduling.

Regardless, it was an important step in my development in the industry. There were so many layers of how exhausted I felt every day. I was almost immediately isolated and disconnected from the rest of the world and my typical in-person, on-site priorities. The pandemic was in full swing by this point, with state masking protocols, grocery lines with five-foot spacers, and newly established hand sanitizer stations.

Professionally, I felt completely alone, with limited to no resources. But I continued to discover new ways to connect with friends and family. Even I got swept up in watching *Tiger King* on Netflix, thanks to TikTok. I explored different attempts to feel connected during this strange period in the world, including revising and sharing indoor workout plans on Instagram and having distant sidewalk coffees with my friends.

As the weeks went on, I started to become more anxious as I recognized the changes happening across the company, with many teams being reduced because of the decline in revenue

---

27 Milestone work is a set point in a schedule where progress is tracked or compared. As an example, project kick-off and final turnover are project milestones.

flow. I worked closely with our head scheduler in Minnesota, along with the assigned trainer, who was in Arizona. They both tried to be encouraging, but it was clear that they had other struggles between family and work priorities.

Close to the end of May, I got the check-in call from my old project executive, Caleb. Honestly, it felt odd. He hadn't given any sort of positive indication about my departure from the project or switch over to scheduling. So, it felt eerie to respond to his questions about how I was holding up. The conversation was brief, with both of us acknowledging the difficulties of this period.

By June 2020, roughly two months after leaving the six-level AT&T Tower commercial tenant renovation, I was let go from Momentum Builders due to budget cuts from the pandemic. To this day, I am not 100 percent sure the pandemic was the actual reason. There were other bits that I considered to be part of the cause, including Kaleigh telling me earlier in the year at my annual review that I was the highest-paid construction manager in the company. I also considered the fact that I didn't just accept the physical and mental abuse dished out by Owen, my site superintendent who threw hard hats and cussed people out when they didn't respond the way he wanted.

When I got the phone call from Kaleigh on June 1, 2020, my heart was racing. At first, I was delighted, since I admired her career and approach to the work we did. She quickly told me it wasn't a social call and let me know that HR was on the line. Immediately, I was told that I was being let go due to the third wave of downsizing taking place. Effective immediately, all devices would be disconnected.

I was shell-shocked. Perhaps in my naivety, I believed I wouldn't be let go due to my diverse experience and background. Reality hit hard.

A few hours later, I got a phone call from Kaleigh's admin. She immediately offered her condolences over the situation and gently coached me through what would happen next. She explained that I would be permitted a one-hour slot the next day to retrieve any personal or imperative items before losing permanent access to my laptop and work phone when all work devices were shut down. We set a time, then scheduled a time for a carrier to come out to collect all work-related equipment (devices, monitor, keyboard). She explained that a collection of documents would also be sent over for me to review and sign. In essence, these documents were types of NDAs[28] that would permit me a termination package of less than a month's pay and sign me up for unemployment.

The next day, I logged on and collected my daily log and other emails that would be good to have on hand in case I had to go to court. There were a few work products I was proud of as well, such as my Procore foreman training and some process documents I'd built. There was so much that I'd accomplished.

---

28 An NDA, or nondisclosure agreement, is a signed contract or document in which you sign away all rights to take the other party involved to court or sue them for any type of damages, neglect, abuse, etc.

But in the blink of an eye, it felt like a few years boiled down to a handful of documents. It was such a harsh feeling. The impact immediately evoked a sense of failure, like somehow, I had done something wrong or deserved to be terminated.

After a few conversations with other women in the industry and upon reviewing my termination package, I decided I wasn't completely confident about signing what they provided. More or less, it kept the contractor I worked for exempt from any fault or responsibility for my firing. In short, I couldn't take them to court if I signed the NDA.

This felt wrong and off-putting. Delaying signing the package and working toward peace of mind, I decided to contact a few labor attorneys. Unfortunately, even with the indication of my harassment and my belief that it was the real reason I was fired, no one was interested in taking it on or even having a consult.

Defeated, I accepted without attempting to negotiate the flimsy package, which included less than a month's pay and no benefits unless I signed up and paid for COBRA health insurance.[29] Not to mention, this was all happening during the pandemic. The whole world was unsure about the future of mankind and what we were socially supposed to do while we were bottled up indoors and away from others with our Zoom calls and Facetime.

---

29 COBRA is complicated and expensive post-employment insurance that offers temporary benefits after group coverage ends.

I took some time to grieve the loss of my role and the potential setback and anguish forced by the whole experience. When I'd moved back, I was so elated and optimistic that I was going to be successful in my endeavors. Instead, I got kicked to the curb and landed squarely on my butt. Mixed emotions ran through my mind, ranging from self-blame—thinking I was the reason the harassment had occurred—all the way to wondering what I should do now.

It took a few conversations with some close friends and allies to quickly learn that I was not alone. I connected with **different people, including a number of other women from the same company who were going through the *exact* same issue.** Those conversations started to ground me and bring me back to reality. And I stopped wallowing in self-pity. A few weeks later, **I made a significant decision that would ultimately change the direction of my life.**

I applied for an MBA program at the Carlson School of Management through the University of Minnesota. It was an idea I had been exploring since my time in Portland, but I'd been a bit fearful and felt very out of my depth. By this point, I didn't have anything to lose and knew that going back into construction was more than likely going to be a repeat of the **same harmful experiences. I did not "fit" or behave as the type of** woman the contractor world told me to be. Exhausted from my repeated attempts to prove myself in the male-dominated space, I pulled the trigger and pursued the best path for reinforcing my expertise: earning an MBA.

PART II

PART II

CHAPTER 8

MBA

*Trusting yourself and allowing the journey to unfold
is the biggest leap of faith you can take.
It's scary, like jumping into a dark void,
unsure of where you'll land. But invest in yourself,
because no one else will.*

The gears were turning and starting to click in place. No matter how hard I worked, showed my skills, or got in the trenches to resolve a problem, I was never going to truly be recognized for my expertise. It was always going to be a dance of how to present the idea or solution in a way to delicately engage the men around me to either own it as though they came up with it or not shut me out during a discussion.

THE TRUTH ABOUT BEING A WOMAN IN CONSTRUCTION

During my time in Portland, Oregon, I quickly realized that many of my male leaders lacked trust in their female colleagues' knowledge. Some of the smartest women would smile too much, giggle, or even pose their thoughts as questions to curate an environment around a decision so the male perspective would feel less intruded on.

This was a rude awakening after living and working in two very progressive cities on the West Coast. I'd truly thought that the men I was around would be advocates and partners. Instead, they undercut and were sometimes downright ruthless in their actions. Thinking of ways to solidify my expertise while expanding on my foundational knowledge boiled down to pursuing my MBA.

It was an incredibly daunting thought, and doubt edged in while I talked to counselors and looked online at programs. Portland had a well-defined sustainability MBA program that excited me, but I was struggling to figure out the details of a schedule that would be flexible enough with my rigid fifty-to-sixty-hour workweek.

The door flew right open once I was let go from Momentum Builders in the summer of 2020. I had started to get the ball rolling with an advisor the previous fall after discussing my idea with a few others who had pursued a similar program.

At this point, there was nothing in my way, so I moved in full force. By September 2020, I was all lined up with the University of Minnesota's part-time and online MBA program and set a pace to graduate in three years. This wasn't going to be a small feat, as I earned fifty-two credits at a full-time pace in a three-year span. The average PhD program is sixty credit hours.

At this point, many of us were still learning to navigate the pandemic, with most places around the Twin Cities were either temporarily closed, had windows for walk-up access, or required masks. U of M had a mask mandate, which some abided by and others disregarded. That rainy fall, I nervously started attending my in-person courses in strategic management and financial accounting.

In the beginning, it was refreshing to figure out a way to connect with folks from all sorts of industries and backgrounds. Learning so much at a rapid pace was a bit overwhelming, but having peers to discuss or collaborate with was rewarding. Much of the curriculum was a mix of outdated materials or was taught by professors who didn't explain or correlate their assignments well with the lectures. This created many unnecessary challenges and headaches. However, I was committed and learned to be nimble while finding additional resources to help me advance my comprehension.

A few months into the program, I decided to find a way to build new experiences and expand some of my skills in new ways. Internships appeared to be a logical next step, providing some financial relief while preparing me to launch in a new direction after I had my MBA. Since losing my job at Momentum Builders in June, I had been living off savings and unemployment, which were quickly dwindling. So I set up a call with my advisor and business school career counselor. They both stated that students in the program typically didn't seek an internship until closer to graduation.

This felt like a nonsensical status quo, especially for someone changing careers. I was dead set on trading my work boots for a leadership position in sustainability. An internship was a sure way to explore those opportunities and hone in on what I wanted to do professionally after graduation.

GAINING EXPERIENCE

After receiving lackluster support from the career center, I started to network like it was my full-time job. I pulled on different threads, cold-calling individuals who had an interesting title or asking friendly past connections to introduce me to new ones. These were some of the harder conversations, and I felt vulnerable as I asked for help while expressing my unsureness about what was next.

I was convinced I was on the right path, and a window did open. I finally got in touch with someone with a local solar installer and utility provider. The person I was able to get in front of just happened to be my soon-to-be boss.

He and I met briefly to discuss some of my interests. Fifteen minutes later, he admitted that he was impressed with my hustle and background. He informed me that his team was looking for an intern, and he thought I would be a perfect fit.

MBA

A week later, I was interviewing for the open slot, along with undergrads and a few ambitious MBAs from all over the country. Other than my scheduling rotation for Momentum Builders and the few online courses I had taken so far, this would be the first time I would be working in a remote environment. It was all very new, slightly overwhelming, and anxiety inducing.

Two weeks later, I was offered the eleven-week internship with a partial promise to extend with NovaGreen Tech. I was downright proud of what I was able to carve out in a short period with limited resources and in a world that felt closed down. May came, and they shipped me a laptop, a monitor, and a few other goodies to get me started. Week one was onboarding, which combined different introductions, a very lengthy welcome from the CEO, and icebreakers to encourage interns from all over the country to connect.

The first few days felt like a dream. People were nice and encouraging. Even working from home started to feel a little bit more comfortable. Except for my manager, my team consisted of two other women. I hit it off with them, as we were close in age, and it was easy to converse and share ideas. I dove into finding new challenges to support, expanding my network, learning from others in different positions of the company, and adding to my sustainability resources.

The first two weeks promised a lot with the team. But over time, I started to see a trend with the older interns, who were like me in that they, too, were changing careers and pursuing higher degrees. We were given less focus and more of a pat on the back, with the direction to figure it out. One of my peers did a great job breaking down some of the recent policies and how they connected to the company, including recent reports from the Intergovernmental Panel on Climate Change (IPCC).[30] It was well done, but no one in the company took it seriously. I had also taken time to develop an onboarding guide to help others know how to use different tools for siting projects. It was a clunky process using Google Earth and PJM (Pennsylvania, New Jersey, Maryland Interconnection) Energy Market data to figure out what plants, transmission lines, and stations were in progress, with their associated stage. I wanted to help smooth the process for efficiency and even worked with a few other business development folks to create and confirm that what I was building made sense.

I soon felt a sense of isolation. I put 100 percent into the little bit they gave me, including researching and reporting brightfields—an intentional way to repurpose abandoned space by redeveloping or remediating contaminated locations. I thought this made way more sense than handing over solar

---

30 Intergovernmental Panel on Climate Change (IPCC) is an incredibly important group of scientists, activists, and environmentalists working for the United Nations. They collect data and share it with key decision-makers across the globe. Check out https://www.ipcc.ch/ for more info.

MBA

leases to already predominantly wealthy farmers,[31] which was NovaGreen Tech's business model. From a strategic standpoint, I created a road map—a solution to grow as a more inclusive company. I provided details, such as better hiring practices and how to promote within and establish a mentorship program. I also made sure to offer a more equal path into leadership for women and minority groups, diversifying the executive team filled with White men of a certain age. This would directly benefit the company in retaining its top talent while bringing more perspective and innovation from new hires.

Halfway through my internship, I met a guest trainer on workplace stress who had an MBA. Excited to meet another woman business owner, I was eager to hear her story. We eventually met offline, and I asked about her experience in getting an MBA and what she felt was most rewarding. Toward the end of the conversation, she mentioned that her husband and his business partner had a consulting firm and might be interested in bringing me on board. They were focused on Lean

---

31 Famers the company was pursuing tended to already be making serious money, in the millions of dollars. This was a missed opportunity to reach a wider community of folks, including information about better access to renewables for disenfranchised communities.

project or Agile planning methods,[32] which were very familiar to me after my construction experience.

I was thrilled to have the conversation. A few weeks later, I connected with her husband, Craig, from the consulting firm. He mentioned what they did and how they made construction run more efficiently and be a better place to work. I was excited at the idea of putting my skills to work and building up a resume. I could see an easy overlap between the sustainability and green practices being applied in a Lean methodology.

Craig said he would talk to his partner, Liam, and circle back. But he also stated that he was intrigued by the idea of bringing me on board. I was thrilled at the opportunity to carve out a path that would positively enhance the trades and construction management itself and create a more inclusive bridge for often forgotten team members. Since I was feeling unsure about NovaGreen Tech and the longevity of my role, this felt like a strong potential next step.

Time went on without much feedback from Craig. Some things had come up, and he was traveling. The weeks of my internship dwindled quickly. I was still seeking mentorship in the industry, and, by this point, it was clear that my manager had checked out. The two women I worked directly with, Alison and Sandra, were brilliant but incredibly busy. My internship

---

32 Agile planning methods are a form of Lean project management. This approach focuses on collaboration and offers flexibility in the schedule. Learn more at PMI.org.

became a bit dreadful as I counted down to my last day with no promise of further opportunity or engagement from others in the company.

I wanted to keep learning about the business but couldn't seem to get anyone to sponsor or mentor me. I was on my own. Toward the end of the internship, I started to grasp at straws. My manager had been so full of promise when he hired me. However, the red flags had increased when he was unable to keep one-on-ones with me, consistently canceled meetings, and even pushed me off on Alison and Sandra like I was their responsibility. The hard fact was that he had overpromised but had no real intention of staying put. He was already making moves. This was the unfortunate hindsight I learned after he quit cold turkey.

Sandra, who was the senior woman on my team, flat out told HR she didn't have the capacity to manage me. The last week, after presenting my capstone[33] project, I continued to check in with folks, hoping for some promise of a position to switch over to. I presented my capstone project, which revised a new business strategy to help align and recruit more inclusive individuals. I hoped that it would showcase how valuable I was to the company and how versatile my skills were.

---

33   A capstone is used as a final marker or milestone to show off what was learned. It's a performance indicator to embody a subject or overall comprehension.

I truly wanted to make a difference, especially since the HR manager had encouraged my honest observation. I was a woman who had worked in many male-dominated spaces and recognized my privilege of being White. So it was important to call out the disparities I observed. I was vividly aware of the disconnect and tone-deaf approach of those I worked with, especially at the leadership level. It was a company made up of mostly White folks, including the CEO, who referenced his Asian wife as a way of understanding minority groups and diversity. The audience I invited to observe from different departments truly appreciated it. HR, not so much.

With three days left of my eleven-week internship, HR reached out to ask if I wanted to work as a construction management intern. I declined, feeling defeated but also insulted. They wanted me to intern for sixteen dollars an hour for something I'd done as a professional with a salary. I decided to take it as a sign from the universe that this wasn't the right fit and that something else would work out.

And I was right.

LEAN PRACTICE

Still determined to get a range of experience and add some income to my bank account, I kept leveraging my network. I had many difficult conversations, including a few with individuals who started warm and then went cold on any conversation about me working with them. Eventually, a few weeks after my internship ended, I finally got a sit-down with Craig, the Lean consulting founder I had connected with midway through my internship with NovaGreen Tech, and his partner, Liam.

Over the next few weeks, we had a series of brief conversations that shaped what my role would look like. It sounded like a dream in many ways, with both of them encouraging my insights and stating I would have the opportunity to build up and shape the consulting firm.

In October 2021, about two months after concluding my time at NovaGreen Tech, I was warmly embraced as the newest team member at Growth Point Advisory—a tiny company with barely ten folks all over. The wife of the principal owners was our admin. It felt sweet and wholesome in a gentle way.

Eager to get to work and with stars in my eyes, I set up my regular calls with Liam, who turned out to be my main point of contact. We checked in once a week when it accommodated both his travel schedule and my courses. He would share some of his project challenges and encourage me to get familiar with the Last Planner principles, a construction methodology paired with Agile planning and a Lean project management approach. Initially, it was a slow start. I was getting familiar with a newish idea of running a project while pulling professional experiences into the mix.

One of the first things I proposed was a technology upgrade to improve our organizational structure and streamline team operations. Everything we did was through Google Mail or Google Drive. At one point, when I asked about a Microsoft license, I was told to just use my school account for work. I firmly acknowledged the company risk of doing such and insisted I get a business account.

After crunching some numbers and doing homework on the current file and communication structure, I estimated an almost $6,000 annual savings and an overall efficiency improvement of 50 percent by switching to Microsoft. However, when I attempted to offer this perspective, including resources to help make it a reality, I was told the company was too small to consider changing the ways it had operated for close to ten years.

That was red flag number one. Here I was being encouraged to be vocal and find new ways to improve our organization model as a business, which promoted efficiency and agility, but was rejected quickly because they didn't want to make the change.

Red flag number two happened close to the end of month one. I was told to stop coming prepared for the team meetings with an agenda. An agenda was a communication tool—a document I used to reference updates on what I was working on or needed help with. Craig and Liam made this request in front of the entire team, suggesting that it inhibited organic conversation. This completely disregarded my need as a new member to feel confident, prepared, and clear about what I was working on.

How I contributed mattered. This felt dismissive and discouraging.

Over time, I started to feel less recognized or valued for my wealth of construction experience. The tasks assigned were a combination of cleaning up Google Drive and building out white paper templates to coordinate with the company's logo and brand. This wasn't what I was hired to do. I offered to travel to certain projects. But they always responded, "Well, we don't know how many people will be on-site or if it will be worth a visit." Sometimes, they said, "Maybe next time."

I brushed it off, since COVID-19 was well into its third or fourth variant. By early 2022, some folks were starting to feel a bit more comfortable in public spaces, but construction projects were still struggling to keep skill trade around and healthy.

Eventually, in mid-January 2022, with snow still on the ground, Liam took me to a project with the familiar range of **grumpy men standing around a schedule, trying to figure out** how to resolve areas they had fallen behind on. Liam walked me around the job site, showing me how they had staged the work and coordinated each scope. Admittedly, I felt like I was being shepherded around and lectured. It was not at all what I'd expected, which was more of a collaborative, shared-experience type of approach. I felt like a child being shown around by their dad. It was a bit infantilizing.

Finally, in early 2022, I was able to attend an all-day training **as a guest with Liam. My role was to be nothing more than a fly** on the wall. Literally, I had no task or hard ask. I was to just sit close to the front and observe. So I did as such. Occasionally, I walked around masked, taking photos of team during lego **play or collaborative planning practice.** At first, it felt nice to be around folks who were interested in learning how to be more agile in their planning and gaining new knowledge. However, it all felt nominal and almost like I was quietly being told to know my place. Even Liam's introduction felt dismissive of my combined work experiences and current MBA endeavors.

A contact I made in the late fall of 2021 ended up presenting my next big opportunity. I had connected with a team working on the Last Planner System, which focuses on software as a service (SaaS). This was an opportunity to fuse much of my construction management and field experience to help shape a project management tool. I also felt like this would be a good growth opportunity since I had no background in developing software. However, as someone familiar with ConstructTech for managing job sites and various scopes, I had a lot of insight to offer.

I found out later that they were hoping to use me as a leverage point at Growth Point Advisory to sell a platform they were bringing to market. I was delighted to learn from them and their level of engagement. At one point, they commented on my skills and what a bonus I would be to their growing team. By early 2022, I was ready for a change and exploring other options.

I was given an application link and a first-round interview with the hiring professional for Bolt Power Tools. The hiring manager was impressed with my resume and background. She even commented on how I was a rare candidate who could tick all the boxes for a role. She went back to my original contact, Afonso, who was leading the Last Planner SaaS tool team. He gave her the green light to hire me.

By March 2022, I was being onboarded with Bolt Power Tools to work on their Last Planner site tool. Energized to start a new adventure and dip my toe into software as a service, I envisioned much calmer waters with greater responsibility. They quickly set me up with all the gear to work remotely and offered nice flexibility with my studies. There was some mention of travel, but since I was more focused on back-end support, it was less of a consideration for me.

Feeling like I had *finally* found a team that wanted me for my skills and insights, I was encouraged that this was the milestone position that would help direct my professional path. What could go wrong?

CHAPTER 9

AUSTIN

Learning how to self-advocate is imperative.
How you let others treat you signals that this is the behavior
you will tolerate. If you saw someone else treating your sister
or best friend poorly, what would you do or say?
That same advice applies to you.
Love, cherish, and protect yourself. You aren't anyone's
punching bag.

After about a week of getting familiar with the Last Planner tool, there was an opportunity for me to connect with my team in person at a customer meeting in Austin. It was a two-day trip. We flew in the night before to attend a client appointment. Day two was supporting a Last Planner event that Bolt Power Tools was sponsoring.

The Last Planner team flew down to Austin to help solidify an ongoing relationship with a main contractor. This was a huge opportunity for the start-up software company to continue to get financial backing. Bolt Power Tools had been struggling over the last few years to gain a robust range of subscribers and sponsors, making it hard to continue the development of the tool. Without financial backing, it was having trouble reinforcing the need for this digital tool in the industry.

In the same period, we were collectively supporting a Last Planner workshop put on by two individuals who were growing in popularity for their very hands-on and real-world approach to applying construction and planning methodologies.

On my first night in Austin, I was sitting on the balcony of the hotel and taking in the spring air. I felt proud to have come so far and to have made my way to a large company. Eventually, the rest of my team arrived with Afonso, including Alex and Karen. Afonso was the lead salesperson of the group with Alex being his right-hand person, and Karen a recent intern turned full-time part of the team. Their job was to sell the platform and build relationships within the North American market. Very tired and a little red-eyed, we all sat together, exchanging pleasantries. Karen was the only other female present. When I went to shake her hand, she stuck it out in an odd way. It was as though I were handling a delicate bird wing.

Afonso, who was the lead for our team, went straight into outlining the next few days. Since I hadn't been given much heads-up, I made sure to take notes on my phone. As I avidly listened and asked questions, Afonso explained the importance of our visit and how it could solidify our relationship with a client, all the while reinforcing the success of our tool.

On day one, I was ready to go. I wore jeans and a blazer with boots and a black trench coat, all designed to make a clear statement that I was comfortable on-site while giving my look just a boost of professionalism.

We all piled into the Uber, masks intact, and listened to Afonso excitedly go on about the client we were meeting and how far we had come in the conversation.

When we arrived at our destination, we were invited to sit in a large, empty conference room with a large screen at one end. Karen went to work, pulling out her laptop and getting her presentation set up. Afonso and Alex engaged with the company CEO, who showed up shortly after we did and effusively thanked us for being there and told us what an honor it was.

The room started to fill up, and I went around, shaking hands and introducing myself. Afonso went to work making an introduction to the Last Planner tool we were presenting. Karen immediately followed, but she began having problems, noting that her laptop was dying. After a quick scramble and a helpful hand from a project manager in the room, we found an acceptable power source and got the large screen to show the presentation.

Afonso eventually turned it over to Karen, who stood and immediately started to swivel her chair with her knee. Dressed in a casual jacket, gray pants, and a T-shirt with tennis shoes, she looked like a young high schooler who was nervous to give their final project. The whole time, she kept talking and moving the chair back and forth.

Doing my best not to be distracted, I focused on the conversations at hand, including wielding my personal experiences on the job site to endorse how useful the tool was and its importance in creating efficient and transparent modes of communication. At times, I even recalled how well the tool captured different details and quickly described the story of the project. Since at this point, I'd had several interactions with the team, had access to a demo, and had even listened to this team in action, I felt like I was a good bystander who offered added value with my perspective.

Yet the foreshadowing of how this experience was going to turn out started to take shape. After the meeting, we hung out a bit, and Afonso commented on my level of interaction. At first, he was complimentary and stated how well I was able to engage. But he added that I was incorrect about some of the information I'd shared. He said I should have sat quietly and let Karen continue to showcase the SaaS tool.

At first, I was a little stunned and asked a few clarifying questions about what I hadn't expressed correctly and how I'd impeded the conversation. I'd witnessed more engagement and interest after a few of the perspectives I provided.

Alex stepped in and did his best to clarify Afonso's point. But, ultimately, both were telling me to sit down and be quiet.

The rest of the day didn't feel much better. We went to a local lunch spot after Afonso raved the entire car ride about how much of a foodie Karen was and how she always found the best spots. Alex talked up Bolt Power Tools, describing how amazing and progressive its technology was and how it continued to be innovative.

The whole time, as I was watching and listening to each of these men discuss and emphasize how flashy and stunning things, people, and places were, I was slowly starting to see behind the curtain. They were trying to overcompensate for their lackluster explanation of the product and service offering. At many points in the conversation during lunch, Afonso talked in circles about the product, almost on repeat. I felt that, in theory, I had a pretty good grasp on what we as a team were offering.

After lunch, we spent the afternoon at another contractor's office, where the Last Planner workshop was being hosted. This was the significant reason we'd flown down under the Bolt Power Tools sponsorship.

The idea I quickly formed for the workshop was that these two men, Kevin and Ken—whom I had heard a few things about—were credible when it came to teaching a relatable approach to implementing Last Planner methodologies. The idea behind this practice was to work from the end of the last scope or before the final turnover of a project. By working backward, the construction management team could build out a better schedule and identify potential gaps or risks. This practice included how to approach or speak to people on your team, build connections with various field folks, and use essential communication tools.

We met up with a videographer and his assistant. They were pulling in large cases of equipment and trying to get a firm grasp of how the area was going to be set up. Afonso kept pacing around, with Alex following close behind and determining which space would be best for video interviews vs. the workshop. They decided on a smaller, less open conference room for the day of the workshop, which was being hosted by two guys who called themselves the Best Builder team.[34] It was a bright room with nice mobile tables and a fair number of chairs. We coordinated with the office clerk to make sure there was an appropriate number of chairs and tables just in case more people attended than predicted. Afonso and Alex were also given a large box containing a bunch of company swag, including little sweatshirt-drawstring backpacks, which we packed with a blue branded notebook, a flat field pencil, and a book detailing Lean building practices.

After about fifteen minutes, there wasn't a ton more we could do to prepare for the day. Afonso led us back out to the large echo chamber, which was the back of the hosted contractor's warehouse, to check on the videographer. Afonso started to

---

34 The Best Builder team consisted of dedicated experts who evangelized the Lean building process.

move the metal tables and flat bar-style stools off to the side so they wouldn't interfere with the backdrop.

The whole time, Karen was off to the side, leaning on a table and seeming unbothered. It was quite comical. I started to pick up a side of a table with Alex when Afonso rushed over and said, "No, no, no. I have this."

There was a cultural difference in this moment that showcased his expectation of women's behaviors, which Karen was appreciating, enjoying, or taking full advantage of.

The afternoon continued to drag on, with Afonso and Alex staying close to the videographer, asking questions, and watching him set up, piece by piece. At a certain point, I attempted to converse with Karen to understand what the full expectations were for these interviews, assuming that, since I was the newest team member, I would be included.

She gave a noncommittal response, saying she wasn't sure but that it would be a mix of talking about the SaaS platform and having the lead workshop hosts talk about their book while promoting the Last Planner tool. I was somewhat excited about this opportunity and feeling included in a big enterprise initiative.

Who knew how wrong I would be until the next day.

Finally, after three hours of spending minimal time contributing to the video and overall workshop setup for the following day, we headed back to the hotel. When we arrived, I was pretty drained. Between the travel time of the evening prior and the day's engagements—including Afonso and Alex inviting themselves to a project site spotted after lunch—I was pooped and ready to retire to my room. But Afonso was emphatic that we had work to do and needed to come together at one of the big central tables near the hotel lobby.

In my mind, we had a full day the next day, engaging with folks and sitting in an all-day workshop. I felt the rest of the day was a time to recharge, not power ahead. This was at four p.m. on a day that had started at six a.m. Without saying a word, I went toward the elevator and hit my floor.

Afonso was a step ahead of me and bluntly asked, "Where are you going?"

I responded, "Up to my room to get my gear."

He said, "Oh, okay," and seemed somewhat relieved.

As I reflect, I wonder what sort of argument would have ensued had I said I was going to my room, period.

I came back down about ten minutes later after chugging a ton of water and refreshing myself. Remember, this was still during the pandemic, and I was cautiously wearing a mask during every outing. I wasn't taking a chance of getting sick with whichever variant of COVID-19 was going around at the time. Between my studies and work, I was stretched thin. There was no room to be sick.

When I arrived back in the lobby, Afonso was still up in his room on a client call. This irritated me a bit, but I still wanted to be a team player, so I kept a positive attitude, opened my laptop, and logged in. I'd been with the team for only a week, so there wasn't a whole lot for me to do other than plug into some of my onboarding training materials.

Listening to Karen and Alex write an email response (which took forty-five minutes between the two of them), I attempted to find a gentle way to add value to the conversation. In their remarks, they were commenting on folks in the field but using terminology that was not inclusive.

I said, "If I may? I would adjust the phrasing to refer to both men and women or use the words *they* or *them*."

Karen and Alex looked at me with blank stares, then continued their conversation. This quickly reinforced how I was *not* part of the team and painted a full picture of what was to come.

Day two of Austin was the Last Planner workshop with Kevin and Ken. It was another early morning with Afonso, Alex, and Karen, and all of us met in the lobby to turn in our room keys. I had made sure to get breakfast in time, unlike Afonso, Alex, and Karen, who appeared to survive on Red Bulls and coffee alone. It was a bit unsettling and left me questioning what it was that kept them all so busy while I was off to the side trying to find ways to support them.

We got in our Uber all masked up. I was wearing the same uniform: jeans, blazer, and boots. Karen was wearing her tennis shoes and jacket, while Afonso and Alex were almost matching in their company zip-up puffer vests. The only difference was that Alex had on his penny loafers while Afonso was in his never-seen-a-site work boots.

We arrived at the contractor's building located in the suburbs of Austin. Afonso led us straight into the moderate-sized conference room for the workshop we were sponsoring for the day. There were two other men there, and they were starting to sift through boxes and storage containers. While setting up for the day, Afonso strutted over and started to shake hands with each of them, excitedly talking about how well the day was going to go and how excited the team was to be there.

Trying to busy myself, I went around and started to neaten up the chairs that were lining the tables and straighten the little gift backpack hoodies on the tables.

Afonso walked up to me and loudly said, "Did you say hello to our workshop hosts?"

I responded, "No, not yet. They're clearly busy, but I will later."

Without a moment's hesitation, Afonso turned around and said to one of the men, who was wearing a baseball cap and glasses, and said, "This is Jessi. She's one of our newest team members."

Mind you, this man was fully bent over and digging through a storage bin.

Turning red and slightly embarrassed for Afonso and myself, I walked over from what I was doing, stuck out my hand, and said, "Nice to meet you."

The man responded, "Nice to meet you. I'm Ken."

Repeating the process, Afonso walked me over to the other host, who had his flowing, mid-length, wavy black hair, and said, "This is our newest team member, Jessi."

Again, I responded and offered my hand. "Nice to meet

you," I said.

With a kind smile, the man said, "Hey, I'm Kevin."

Afonso then moved on to make small talk with Alex while I proceeded to go back to what I'd been doing. Karen was off in the corner on her phone, half watching the room of people and the activity.

During the moments that passed after this awkward and forced introduction, I felt slightly betrayed—almost infantilized or not trusted to know when the appropriate time was to introduce myself. Afonso had acted like I wasn't a professional. Like I was clueless and needed special guidance. It was a bit frustrating, since I had well established my background and skills before being hired, long before day two in Austin, Texas. Plus, I thought it was downright rude to interrupt people who were focused on trying to get equipment set up for their day of activities.

Regardless, the morning went on, the videographer showed up, and people started to trickle in.

Bathroom Selfie

The two individuals hosting the workshop, Kevin and Ken, were well-established construction professionals who went around evangelizing about the need to use the Last Planner system on job sites. They had a fun and somewhat humble approach to their workshop, which was pleasant to observe, albeit still very male-oriented.

Early on, there was a callout for our company, Bolt Power Tools, as a proud sponsor. At that moment, the shout-out was conveyed to Afonso, Alex, and Karen. It stung a bit. Yes, as a new member of the team, this quickly made me feel a bit off-put. This gap continued to grow throughout the day.

As the morning session continued and the whiteboards started to fill with notes and pages of examples, we were nearing our first big break of the day. This meant mingling and introducing the Last Planner tool that our company had built and was offering. Immediately, people walked over and started to introduce themselves to my team, who didn't include me in the conversations. Even though I was sitting next to Afonso, Alex, and Karen, many people elbowed past me to get to the individuals who were wearing clothing that had the Bolt Power Tool logos on them. Not only that, but in each conversation, I watched the three amigos not even attempt to introduce me.

So, instead of being a self-pitying martyr, I thought, *Fuck it. Let's go meet some folks and learn something new.* I scanned the room and, without a single business card in hand, found a few women who happened to be project managers. I asked them what excited them about the event. I inquired if there was a digital tool that could work similarly and whether they could see their companies and teams using it. To this day, I am still connected to a few of those very women.

After about fifteen minutes, I returned to my seat again and sat next to my team.

Time ticked on, and lunch approached a little before noon. Everyone broke to grab some yummy catered barbecue and sit among folks. Still COVID-conscious and, by this point, pretty ticked off, I decided I needed a bit of quiet time. I sat alone outside on a curb, reflecting on the morning. At this point, I had collected a fair number of business cards from folks from construction technology companies, general contractors, design firms, and subcontractors. I felt proud of having reframed my experience and the distress my team's exclusion had caused by instead polishing up my networking skills.

After lunch, the workshop continued with a brief section during which Afonso, Alex, and Karen stood up front to showcase the Last Planner tool our team supported. It was clear that they were very proud of the work they did.

The group came back and sat down as the two main workshop presenters picked up where they'd left off in the training.

Moments after sitting down, Afonso loudly whispered to Karen, "Did Kevin and Ken wrap up their interviews over lunch?"

Karen replied, "I think so."

Afonso turned to Alex. "We three should go back there and do our interviews now."

Alex agreed.

So, while the workshop was going full steam, Afonso, Alex, and Karen stood up in the back of the room, walked toward the front, and then went out the side door.

Slightly confused and unsure why I wasn't involved in this process, I started to piece a few things together.

The videographer had spent all that time the day before setting up to have folks be brought in to discuss the digital Last Planner platform. I'd been again left out of Afonso's plan while Karen and Alex were privy to it.

I was borderline livid at this point. How much ruder could they possibly be? They'd sponsored an event with the two hosts, who were in the middle of giving their workshop, and the best thing they could do was have a full-blown conversation and then leave in the middle of it. I quickly realized that all my inquiries to Karen the evening before had resulted in me not having anything to do with these interviews. Again, as the newer team member, I felt completely excluded and dismissed from this *entire* process.

Had I known before landing in Texas what the agenda and expectations were and what my role would look like, I would have been less shocked by the treatment and would likely have decided against coming on the trip. Between the roller coaster and the confusion, along with Afonso's damsel-in-distress treatment, I couldn't get out of that place fast enough.

I was fortunate enough to have already arranged for a friend who lived in Houston to drive over to get me.

That afternoon, around two thirty p.m., kismet or fate wrapped up the event, and I was out.

CHAPTER 10

PIVOT

*Pivoting into a new career can be the most freeing
and scariest leap. Self-investing means trusting yourself.
It means saying, "I can, and I will."
It means not accepting it when someone says,
"No, you can't."*

The sad part was that this early experience gave way to the beginning of worse ones. Afonso became controlling, attempted to be manipulative, and even harassed me to the point where I would get nervous on calls with him. He once had me create a call list using Excel. I, the construction management professional who was getting her MBA—the same one he'd been thrilled to pull from a Lean project firm—had been put to work making a call list.

What he attempted to do with my skills didn't last long. I'd soon networked with enough folks and found many meaningful ways to contribute. One of my favorite projects while supporting the SaaS team consisted of offering user feedback to the UX[35] design team on global calls. The designers were stuck on how real-time users on a job site would use the tool. This included Procore integration.

I was also fortunate enough to work with our amazing marketing team to build tangible metrics that tracked the success of the European and North American teams.

Despite my best efforts, Afonso continued to try to track every one of my movements. He would even tell others I met with that I wasn't allowed to work on their project. When I had check-ins with him, he would ask why I was working on any given item. He never had anything specific to give me.

Finally, it happened. The breaking point.

I had been doing my best over roughly five months to be a team player. I tried to learn as much as possible about construction tech and SaaS, and I worked on diverse international teams. My contribution included offering a seasoned construction and trade perspective.

However, Afonso frequently failed to embrace my efforts. On several occasions, he got defensive and even yelled. I got to the point where I would shut down in a meeting or be quiet.

---

35 UX/UI designers create user-centric experiences. They focus on meeting the needs of the target market and building a solution that serves them.

One of the more significant items I attempted to offer support on was our sales approach and pitch. In that project, I pulled directly from the work I did as a kid with my dad. After sitting in on our in-person pitch (refer to Austin) and a good handful of virtual sessions, I had a pretty decent observation to offer.

I recognized that little things could make a significant improvement, such as using a call script to help with time, starting with the most relevant information, and keeping it simple.

But none of this was well received, and I was immediately told that I didn't know what I was talking about. I was asked how I could dare question the way they connected with their customers. Interestingly enough, about a month later, our newest salesperson, who had well over twenty years of selling success in Germany, proposed the *exact same thing*.

Seeming tired of my feedback, Afonso put me on the spot and said he wanted me to give a sales pitch. He asked how long it would take for me to be prepped and ready.

I said two weeks and put a day and time on the calendar.

In the two weeks that I prepped, I adjusted several items. Ditching their deck and starting fresh, with very simple, straightforward slides, I used a storytelling format and personal experience to capture the audience's attention. Most of the time, I referenced a script to help with time and flow. Once I felt like the visuals and info were in place, I scheduled one-on-ones with a mix of folks in sales and marketing who had no idea what product I was pitching. I took the feedback and refined the deck, but I received overwhelmingly positive remarks on what I had pulled together.

The day of the presentation, I made sure to include important names, including our big boss over in Germany. When I'd just started to dive in, Afonso interrupted me and started to ask off-the-cuff questions.

I calmly switched over and either answered his questions or flipped over to another slide as a reference point. To say the least, it was *brutal* and *abusive*. Not only that, but Afonso flat out hung up on me fifteen minutes into the presentation. Later, he sent me an email saying he wanted to review the session with me and tell me how well I did.

I was stunned at how poorly managed this review session was and how I seemed to be intentionally set up to fail. As soon as I hopped on the virtual meeting, I knew something was off by the way Alex and Karen were behaving. Both seemed almost distracted and were quieter than usual. Afonso started out by **laying into the flow of the information, pointing out that I didn't** understand the tool and wasn't able to jump from explanation or inquiry quickly. He said I was lacking in my comprehension and that I should spend more time using and understanding the Last Planner tool. Afonso also commented that I didn't try to book a follow-up call with him or the team.

I responded to his critiques by pushing back with the instructions I'd been given to provide a sales pitch, not be tested on a sales call in a given customer scenario. I reinforced this point by using Afonso's own words against him: he'd instructed me to provide *my* suggested version of a sales pitch. This assignment was supposed to involve me taking all the inputs and insights I'd accumulated over the months I'd worked with the team to provide a fresh take on how to engage the customer.

Alex and Karen continued to sit silently. It reminded me of two kids sitting and staring at their shoes. Afonso made a few other comments about timing and said I'd been too quick to go through content. He claimed that I didn't cover enough context about the product or the team.

Again, I repeated myself, stating that this was meant to be my proposal of a sales pitch for the SaaS Last Planner tool, not a scenario for teaching me how to sell the product.

After about twenty minutes of verbal lashing, Afonso finally asked how I thought I did.

This I was ready for with my notes. I told him I felt prepared with a script, which, even though I didn't necessarily read off it, kept me on time and focused. The slides were minimal and not busy. They kept the listener focused on the information I was sharing. The final point I drove home was that the personal experiences made the information relevant for the audience, a.k.a. the client.

Afonso's face got red, and his eyes bulged a bit with each comment. To add a final pressure point, I informed him that I had tested the presentation in front of others, including seasoned sales professionals, who'd given me positive reinforcement and even stated that my presentation was in line with what they themselves practiced.

By this point, I tuned Afonso out and let him rant about my inexperience and about how I didn't know what the customer needed. Let me just say again that I have a background as a welder, and I've worked in the trades and as a construction manager overseeing different scopes.

But no, I don't have a clue what the customer needs.

It was clear that I had no one to back me up or tell Afonso that he needed to adjust his behavior. Fortunately, I had learned some serious networking skills early on in my career and had forged a few connections throughout the company. One of the incredible women I had a great relationship with knew of some of the challenges I was facing and how defeated I was. She connected me to another woman, E, who was looking for someone with my skills for her team. About a week later, I met with E, who turned out to be a significant person in my life even to this day.

PIVOTAL MOMENT

E and I sat down to learn about each other's stories. I told her a bit about where I was, between studying for my MBA (at this point, I was in my second-to-last semester) and what I was doing in the power tools division. After some engaging conversation and laughter, E commented on how impressed she was with my background and skills. She bluntly told me that her team was looking for someone with my talents.

I was a bit shocked and honored at the same time. Here I was, sitting in an initial conversation, and somehow, I had made such an impression on this amazing woman that now she wanted me on her team. Emphatically, I told her to sign me up!

We walked through some of the potential challenges and how to navigate them. First was how to address the disengagement from my current team and manager, Afonso. Second, I had to get the application in the system, then processed. We agreed to meet again in about a week to see where things were and to give E some time on the back end to address any potential challenges.

A week later, she and I chatted. The update indicated that everything should be smooth sailing. I needed to let Afonso know as a common courtesy, but from a permissions standpoint, there was no system barrier to prevent the changeover.

E suggested that after she created the application, I should apply right away and talk to Afonso to inform him of the change. She also told me that she'd had a conversation with HR to let them know of the other issues that had been occurring on the Last Planner tool team.

It took her a few weeks to get the application created for the role. I applied and then set up a call with Afonso.

Sitting at a coffee shop late one fall morning, I shakily responded over the phone to a beating of questions from Afonso. He went on and on about how he was there to help and that he could have gotten me any role I wanted. He asked, "Why do you want to move over to this team so bad?"

I responded, "Because it lands me squarely within the renewable energy industry, which is exactly why I am pursuing my MBA."

For thirty minutes, Afonso continued to quiz me on why I didn't come to him in the first place and how he had a solid reputation throughout the company.

The call ended, and I took a deep breath, then quickly messaged E to let her know I'd given Afonso the heads-up. All we had to do now was wait a week to see if any other contenders came up for the role I'd applied for with the new venture-building team.

In early October 2022, I started working with the Venture Building Group, focused on solid oxide fuel cells. To say I was thrilled is an understatement. This was a whole new chapter and challenge since I had never worked with and had limited knowledge about SOFCs.[36] It was a ragtag group of nine of us in total in the States. The rest of the team, which numbered in the hundreds, was back in Germany and Japan and focused on engineering and deployment.

It was an exciting time to be a part of the clean energy transition, and the timing couldn't have been more perfect, as I was nearing completion of my MBA.

---

36 Solid oxide fuel cells (SOFCs) offer a specific process of capturing gasses and transforming them with the use of electrical current and cathode materials. Think of lots of ceramic and electrical current in a pretty steel box. Bloom Energy. (2022, May 19). Everything you need to know about solid oxide fuel cells. https://www.bloomenergy.com/blog/everything-you-need-to-know-about-solid-oxide-fuel-cells/

I quickly got acquainted with my new team, especially the leads, who were Max and E. We had a series of priorities to help us understand our target customer—the hydrogen market—as it existed and was expanding in the U.S. and how to message appropriately. A big part of my role was to build robust research around the potential clients and competitors. This included setting Google Alerts, paying attention to market trends, combing through and participating in customer interviews, and reviewing current or upcoming hydrogen hub funding.

The work we were doing was complex but rewarding. It felt like together, we all formed a super brain of knowledge and expertise that was focused on understanding the product we were selling and then learning how to get it in front of the best customer.

Energy markets are diverse, with varying nuances that get broken down based on regional and municipal differences. Our research painted an incredible picture of what our customers needed, but it also told the truth: that our product was too expensive and too slow to market. Many of our competitors had already been doing this for more than ten years, were scaling up, and had many investors lining up.

My team functioned very well. The credit goes to E and Max because of their ability to listen, engage, and, most importantly, be empathetic. E and I had regular check-ins, during which I would offer updates while sharing some of my challenges. In return, she would give key advice and suggestions for improvement. She was the first leader I'd worked with in over five years who made me feel capable and confident. When I joined her team, I felt part of it from day one.

But our team wasn't perfect. It had plenty of bumps along the way.

For instance, our executive sponsor, Olin, who reported out to the C-suite, had an approach that was very much the opposite of the one E led with. He was coarse and curt. If he wanted something, he would ask bluntly. And if you didn't get it finished, that was your problem, not his. Every so often, we would have report-outs to him, and they were not always positive reinforcement for the work we were doing. I struggled to connect with him even on a one-on-one basis.

When I set up a call with him early on after joining, I messaged him the dates I was available. Technology ended up failing me, screwing up the invite, and causing the meeting to be scheduled when I was out of the office. This resulted in me being a no-show, which didn't feel or look good. Well, Olin felt it was necessary to send a very passive-aggressive email that included a smiley face. Even though I had an away message and my calendar was blocked out, this was my fault. But it all worked out. I rescheduled and made my second impression impressionable.

Later, when I talked to E about the ordeal, she was encouraging and said, "There's always another opportunity to make a new impression." Over time, bits and pieces fell into place, and I started to understand her challenges with Olin and how she'd decided to overcome them. He wasn't going to change, and neither was his role. I took it as another opportunity to learn what *not* to do as a leader.

Time flew by, and I eventually got closer to the end of my studies. We all got together in Ann Arbor, Michigan, for a two-day team training exercise. It was so good to see all those virtual faces in person and to finally give E a real hug. On the first evening, we got to travel out to the University of Michigan and spend time in the football stadium. We also enjoyed conversation with some yummy food.

The next day, we spent the entire time in a conference room as a team, working through our mission, vision, and other challenges. It was a series of unique exercises that encouraged us to re-envision how we operated as a team. Lots of sticky notes, perspectives, and stories were shared. By the end of the day, we felt like we had made progress, but we were tired. We ventured off to a local brewery that was focused on ciders. It was a quaint spot and included a tour. We played some games and went to dinner. Eventually, we all made our way back to our hotel rooms to rest up for the next day and the conclusion of our workshop.

On day two, we pulled all the ideas, rough images, and sticky notes into more succinct, clear statements. It was quite an accomplishment but also a bonding moment as we watched these fragmented understandings come to life. Toward the end, I became emotional when I realized that this was probably the last time I was going to be in the same room as that team. I expressed my appreciation for the connection, empathy, and consideration we had for each other. There was an admiration and respect for one another that I hadn't seen in a long time—not since my time as an undergrad at Ball State University.

Our workshop advisors wrapped things up and established the next steps, recognizing that many of us had flights to catch. Lots of hugs and words of appreciation were passed on as we said, "See you later."

CHAPTER 11

DIRECTOR

I did it!

January 2023 became a whirlwind of a month. I connected with Natalie, a fellow MBA student from the Carlson School of Management at the University of Minnesota who worked for Juniper Core, a large multibillion-dollar manufacturing company. Her role was focused on sustainable infrastructure. She and I related on many levels when it came to the various challenges women face in male-dominated industries. Natalie gave me the download on the ups and downs of her more than ten years with Juniper Core, Inc.

She was optimistic about most of her experience and the flexibility she had to grow. At the same time, she commented that the company desperately needed more women and minority folks who could redefine the perspective and change who was at the top. Like in most large American corporations, executive leadership looked and acted very much the same—old, White, and male.

We had a series of conversations, and she eventually mentioned a role on her team that was opening. It piqued my interest, but I didn't want to go in blind. So I asked if she could refer me to the person who'd left the business operations position in question. She made the LinkedIn introduction to that person—a woman named Emma—and a week later, she and I were on the phone talking about what she was doing now and what she'd left behind. Over time, I gathered more insights and started to understand the business operations responsibilities I was potentially stepping into. I prepared myself for an intense **role that was going to look drastically different from the tight-knit** team I was with at Bolt Venture Group.

I applied to the position, with the generous inclusion of both women—Natalie and Emma—as references. Within a week, I had an email asking what my availability was for a screening call with the recruiter.

My heart was *pounding* so hard that I thought it was going to jump out of my chest. But when I sent in my info and applied, I dismissed any notion of the company considering me as a serious candidate.

Up to this point, I had struck out quite a bit with three **different companies I had applied** to. I was feeling a bit overwhelmed as I neared graduation and the knowledge that I didn't have a continuing position with Bolt Venture Group. The team I adored didn't have the financial capacity to keep everyone since the solid oxide fuel cell technology we were supporting still wasn't ready for mass production.

Without delay, I set up a call with the recruiter at Juniper Core and went through the screening process. I came prepared with my expectations and questions in hand. It was a delightful call, which was encouraging and gave me hope for a good fit.

A few days later, I received another email asking for my interview date availability. I quickly sent over what I could conjure up and accommodate with my course schedule as I finished up the last leg of my MBA.

During round one of the interviews, I was so nervous that my palms were sweating as I turned on my camera for the virtual call. Wearing my power colors—a pale pink turtleneck and tan plaid blazer with hints of baby blue—I was dressed feminine but professional. The first thing I noticed was that the interview panel consisted of one male. The rest were women. One person kept their camera off the whole time as they chased their toddler around the house (as they indicated early in the interview).

The panel asked a wide variety of questions, and I answered honestly, pulling from experiences and taking every opportunity to turn my answers into a story. My approach was to take my responses and position them from an empathetic yet firmly collaborative team angle. This is who I am, and I wanted that to be front and center.

In the end, I made sure to thank them and bluntly commented that I appreciated how many women were on the interview panel. It wasn't something I saw too often.

A few weeks passed by, and I figured I'd messed something up or wasn't who they were looking for as a candidate. I touched base with the recruiter, and she informed me that it was the opposite—the team was thrilled with me. But, due to a series of events and vacations, coordinating a second round of interviews was a challenge.

I was on pins and needles for the better part of three weeks. At this point, *I wanted this job*. It was going to be a great addition to my resume—not to mention a significant salary increase and solid benefits. Plus, it was *the* spot to pivot into sustainability using my MBA.

Round two was finally scheduled for the end of February. I was meeting with the rest of the global team, including the vice president.

This was it. I had made it to the last few candidates.

Repeating my same efforts as the last time, I showed up looking polished and confident in a black Scandinavian sweater. I kept my answers heartfelt and used the same storytelling approach. The vice president was the loudest in the conversation, showing up late and leaving early. He asked less specific work questions and more general ones, such as "Do you like building PowerPoints?" and "How do you convey messages to bigger teams?" This was the shortest of the interviews, but I made sure to thread in very strategic pieces from their sustainable offerings to showcase how I was paying attention.

To set myself apart from other candidates, I pointed out a challenge and proposed a solution. The challenge was something I'd taken note of from the first interview and found in the homework I'd done to investigate the company's public-facing info. And that was the lack of marketing material about the specific team I was interviewing for. For this team to build transparent credibility, I offered a solution: develop a more streamlined pipeline of content and then utilize and leverage internal team members to market.

This raised the vice president's eyebrow, and he chuckled. After twenty-five minutes, we all said our goodbyes.

Two weeks went by. Then the vice president sent over an email asking if I could come down to Tennessee at the last minute. Unfortunately, that wasn't an option due to my class and work schedule. Soon after that, I got *the call*. I was offered the role of global director of business operations for sustainable infrastructure.

I did it! I finally crossed the threshold I had worked so hard to accomplish since my undergrad years. I'd obtained a position that hit every part of my experiences and skill sets while expanding a global team. This role was going to enable me to bring forward my passion for addressing climate change through decarbonization and threading the technical details into a strategic visual story, emphasizing my creative communication style. I was beyond excited.

It took a week to get the details dialed in for my offer and to negotiate PTO, a bonus, and my salary based on a competing offer. Finally, these were locked in, along with a starting date that gave me time to graduate with the MBA and conclude my time with Bolt Venture Group.

Once the offer letter was signed, I contacted my current manager, E, and let her know the exciting and sad news. There was no surprise, as I had let her know that I was actively looking for a role since our current team was struggling and there wasn't a permanent place for me. The technology we were supporting still wasn't ready for mass production, which meant there wasn't long-term funding for our venture-building team. As much as I enjoyed learning about and enhancing my insights into the hydrogen market and solid oxide fuel cell technology, there was no opportunity for me to stay beyond graduation. She was elated to hear my new title.

We mapped out what my turnover would look like and what my final day with the venture-building group would be. I was taking off a few weeks between graduation and the start of my new role. As bittersweet as it was, I am forever grateful for her kind leadership and her willingness to bring me over to her team. I was in a bad spot and not being used to the fullest extent of my skills. The world needs more Es in it.

DIRECTOR

Official MBA Grad

Standing at the heart of Carlson School of Management

Trying not to cry as my mom takes a photo

THE TRUTH ABOUT BEING A WOMAN IN CONSTRUCTION

On April 11, 2023, I started my shiny new global director of business operations position at Juniper Core. I easily spoke to over one hundred different professionals in the two and a half years leading up to my graduation in the spring of 2023. To say I was scared and worried that I wasn't cut out for the role is an understatement. This was a fully virtual role with some promise of international and local travel.

The first week was the typical onboarding. The chief of staff was my initial point of contact on my first morning and gave me pointers on where to start and what to get familiar with. Over time, my schedule started to fill out with meetings and one-on-ones in which I introduced myself to team members. Most were very welcoming, some were skeptical, and others helpful.

Personalities started to take shape, and I was soon able to identify the louder voices in the room. I was also able to determine which folks got work done and which made noise but would pass the baton to others to deal with.

The politics on this team was much more amplified compared to prior experiences. People started to reach out to ask for help, which I was happy to offer. Some of it consisted of PowerPoint presentation materials, while some was made of marketing content team support and questions about how to boost the sustainable infrastructure brand. It was an interesting mix, but at times, it was confusing and a bit challenging to navigate. We had teams that had expertise in these subjects. But they were so maxed out on various projects or didn't have a full enough understanding of what our team did that the end result lacked clarity or relevance for customers.

One of the first needs I identified was a connection between resources and teams in the field. We had branches focused on sustainable infrastructure. But with changes in content—such as white papers, training, and market reach—there were things they needed better access to. I started to reshape one of our biweekly calls to engage specific, timely updates to communicate these changes to the field. In addition, I helped add more updates on our team's group channel and included our top three languages (English, French Canadian, and Mexican Spanish) to ensure teams were being fed timely updates they could use with customers in their markets.

As time progressed, more pain points were identified. Collectively, I started to gauge gaps and disconnects across the team. Some of the major elements included a lack of understanding of sustainable infrastructure, which is a complex topic. This appeared true both inside the company and among most customers. I worked with our public relations manager and global marketing lead to create a plan to boost our global team profile, which included an internal podcast. This took care of two needs: creating a field resource and amplifying the sustainable infrastructure global team to the greater company audience.

Sustainable infrastructure is the ability to take an existing or new building and make it healthier and greener. Our company's approach was focused on helping our customers reach their sustainability goals. This could range from upgrading or replacing lighting systems and adding data collection capabilities to track how well a building is performing all the way to working with key partners to offer solar to minimize a building's energy footprint. Ultimately, our division's goal is to mitigate the greenhouse emissions most industries make up. Whether a client is a manufacturer, hotel chain, or data center, a significant energy footprint is needed to maintain operations. This is where we can come in to design an efficiency plan to make the building systems more economical and environmentally friendly.

Although our team was a member of our leadership business resource group focused on creating a global understanding of sustainability, many still didn't know that we existed. Bridging that awareness was key. I also started to work closely with our partner's advocate and workforce development team, in conjunction with learning and development staff, to build out relevant training for the field. The focus was to tune into or repurpose content for the company's local branch needs.

None of this was seamless, and it was all incredibly difficult, with many bumps and much resistance. A lot of what I'd thought was relevant was sidetracked by self-promoting efforts or the need to shed light more specifically on the partners we worked with instead. My role continued to evolve to be less and less about operations and focused more on messaging, branding, and communications.

Determined to go beyond those three things, I found ways to plug into bioenergy, circular economy, and even mid-market scaling. These provided a new perspective on the business and the company's operations. I continued to take on more responsibility and move away from building PowerPoint decks for folks. After a few months, I was told that what I was creating wasn't what people wanted. This was disappointing since I have always been good at building decks, but I also had more context and took a collaborative approach. The idea that I was supposed to take a scribble or kindergarten sketch and build it into a polished deck was unreasonable.

I also noticed a shift as my vice president's check-ins became less consistent. When we did meet, his asks were more in line with wanting me to boost his profile by creating an Instagram account and Facebook page or updating his LinkedIn profile. These were more of a self-promoting thing and had less focus on global sustainable infrastructure business. Over time, I was feeling more like his admin, doing "woman's work" instead of contributing to strategy or bringing a field perspective.

Very little of what I was doing had anything to do with my background in construction. The MBA was serving me well in working with cross-cultural management and diverse international teams. I did as I was asked while still focusing on other things and soaking in as much as I could. The funny part about all of this was that the vice president I was working under had a chief of staff and an executive assistant.

Over time, I tried to find ways to build a more inclusive culture. In August, I teamed up with a vice president who was local to Minneapolis to bring remote workers together. We set up a monthly cadence with a small group of team members from all sorts of divisions. I joined all the business resource groups. Even though I was not able to attend everything, I at least got a pulse on what folks were doing. My main attention and focus were on our sustainability and women's business resource groups. Both are imperative in expanding important messaging and resources around hard conversations. I even spoke to a few folks about our recruiting practices, pointing out that we were missing the mark after I was told, "Women don't like to crawl over or under things that will make them get dirty."

My comment as a former welder and construction manager was, "Have you tried working with women in STEM groups or HBCUs?"

The response was "Good food for thought. But our practices are very inclusive."

PIVOT HIGHER

Time flew, and by October, I was starting to get nervous. Since I'd joined in April, there had been a significant series of firings or, as the company called them, riffs. A serious incident had forced budgets to be slimmed up and refocused. I had watched talented women around me get fired. And we were a company that had a women-to-men ratio that was significantly below the market. We were at 24 percent, whereas most similar industries were closer to 29 percent.[37]

With my role being quite fluid, I was worried about being on the chopping block. Plus, I wasn't happy and hadn't been for

---

37 Women in the workforce facts and statistics - Glassdoor career guides. (June 6, 2021). https://www.glassdoor.com/blog/guide/women-in-the-workforce-statistics/

a significant period. What I'd thought I was going to be doing for the greater part of the team and business had become more about promoting the vice president. My one-on-ones became about his needs, not an actual check-in about how I was doing or where I could improve.

I had started a team series called Meet So-and-So. It allowed me to interview a diverse range of team members and help promote them through our North American email newsletter. One individual was an executive director for Operations, and he and I hit it off. At one point, I bluntly asked if we could set up a biweekly touch-base.

After a few of these, he asked me if I would be interested in **starting up a project/program management office.** This is a fancy term that means identifying challenges across a team and then **building efficiency and synergy.**

I was thrilled and gleefully said yes. But I added that I would like more details. We talked about it a bit more, ideating, and he said he needed to do some more on his end to put it into action.

My final check-in with the vice president of the Global team was days before Christmas 2023. I was to move over to the North American team, pivoting inside the company, while still focusing on sustainable infrastructure. Only now, I'd be director of strategic initiatives. When I completed my turnover, which took a month, the role I was leaving behind encompassed the work of about six people. During the turnover, I designated or **made offers to individuals to take over specific aspects of what I** had been working on.

THE TRUTH ABOUT BEING A WOMAN IN CONSTRUCTION

By New Year 2024, I was officially focused on improving operations to help drive efficiency for our teams so they could work optimally toward the clean energy transition. That is where I sit and serve today, focused on amplifying and elevating field voices, addressing challenges through collaboration, and working as a change maker to help solve the climate crisis.

CHAPTER 12

CLOSING

"Unsure to limitless—trusting myself again, building a path I want and deserve."
— Jessica Lynn, MBA

My journey has been a series of bumps with twists and turns. My background as a welder keeps me humble and makes me a better advocate for the people who are typically forgotten. As a construction manager, I have the project management skills needed to organize and take things from vision to reality. The MBA takes all those skills and places a business lens on them, providing strategy and decisive action. It also puts behind my name a magical authority as a woman in a world operated by and made for men.

Working with Bolt Power Tools, I felt the continued dismissal by male authority, but I learned the power of networking and didn't just accept this treatment. This challenge improved my advocacy skills so I can better help others today. Juniper Core offered a lesson on identifying microaggressions, recognizing what a toxic work environment looks like, and knowing how to navigate. These experiences serve as an ongoing lesson on how Corporate America operates and will take everything it can from individuals regardless of the cost or toll. It is up to the individual to flag and set boundaries early on.

But also lean into what you can learn, and know what resources are there to help you grow and adapt. Those will be useful in the future.

I am proud of all my accomplishments, and I strive to bridge resources and elevate the voices of those typically dismissed. As a community leader, hype woman, and change maker, I remain committed to making the world a better place by setting an example and throwing elbows to make room for others wherever I can.

Using my resources and privilege, I started a company called Pivot HIGHer (Hire Her) Plus with a team of brilliant women to extend those much-needed resources for career changers, skilled trade workers, and green job seekers. I've had to pivot or change careers multiple times to find a better path, so this mission is at the heart of my personal experiences. Research reinforces this shared challenge that many women, nonbinary, and minority groups experience.

CLOSING

The clean energy transition has also amplified the challenges many face as they try to mirror their skills into sustainability roles. As a team, we are working to bridge the gap for women, young girls, and non-binary who may not know of other paths to success such as trades, financial literacy, and mentorship.

Together, we can make this world a better, brighter place.

AFTERWORD

AMBITIOUS WOMEN PAY THE BIGGEST PRICE

Be ambitious, but not too ambitious.
Communicate effectively, but don't be too direct.
Be a team player, but don't outperform the men.

Jessi Lynn is one resilient motherf*cker. She literally has "bootstrapped" her way through life—from being a welder, construction manager, global director, and now an entrepreneur—navigating not only the barriers of growing up in an impoverished home in the Midwest but also being a woman in a culture that has been designed and built for men to succeed.

THE TRUTH ABOUT BEING A WOMAN IN CONSTRUCTION

She is driven. She is determined. She has grit. She is ambitious. But ambitious women tend to pay the biggest price in our society because they challenge the status quo and aren't afraid to do it. It becomes this catch-22, if you will, where you're damned if you're ambitious and you're damned if you're not.

Her story and what she's had to navigate reflects this. And though Jessi's journey is uniquely hers, her experiences and the behaviors exhibited by others toward her, especially by men, are relatable to many women working in historically male-dominated industries. The punishment she received by speaking up or being direct, the harassment and abuse, the undermining. As Elizabeth Eting, a global CEO and entrepreneur, wrote in a *Forbes* article titled, "The High Cost of Ambition: Why Women are Held Back for Thinking Big," "The things men are praised for—assertive action, commitment to principle, lofty goals, refusal to compromise—are often the very things women are penalized for."[38]

As an ambitious woman myself, I've faced a myriad of situations, especially in the workplace, that mirror Jessi's. I may have grown up in a different part of the country and navigated a completely different industry, but reading her story I'm thrown

---

38 Liz Eting, "The High Cost Of Ambition: Why Women Are Held Back For Thinking Big," Forbes.com, April 24, 2017, https://www.forbes.com/sites/lizelting/2017/04/24/the-high-cost-of-ambition-why-women-are-held-back-for-thinking-big/.

back in time, reliving similar feelings of anguish, frustration, and sadness that I've buried deep. Because that's what we've been trained to do: bury it and play by the bullshit rules or leave.

But if we keep burying it, how as a society are we able to get better? We need to dig it up. We need to talk about it. We need to hold space for it. That's what I love about Jessi putting her lived experience on paper. Her voice gives others both in and outside the construction industry the courage to speak their truth and shine light into the dark areas of our cultural consciousness that have been operating on autopilot for way too long. Her story is a call to action for all of us to envision and work toward a future that values and uplifts every individual, regardless of gender.

Vision for the Future

I believe so much in the skilled trades and the construction industry being a pathway to a better life especially to those who were dealt a crappy hand of cards. It's an engine of economic mobility, a path to financial freedom. Even working with your hands is said to alter our brain chemistry, opening new neural pathways and strengthening existing ones. A whole slew of opportunities emerges, sparks of possibilities. Things that didn't make sense begin to. Doors that were closed, open. Jessi is living proof of that.

But there are real cultural barriers that exist in the industry that deter diverse talent from coming in or divert diverse and ambitious talent out. And though the number of women entering construction has increased slightly over the past several years, right now, women make up 11 percent of the industry overall and only 4 percent of workers in the skilled trades.[39]

With a huge influx of people aging out of this workforce in the next decade, compounded with the climate emergency and the U.S.'s aging infrastructure, there is skyrocketing demand. We are in desperate need of new blood entering the industry, and it's a huge opportunity for the industry to do things differently that drives more equity into our society. But the challenge isn't only attracting talent, it's retaining talent.

So what do we do?

Industry leaders and the C-suite, it starts with you. It's not enough to slap "DEI" in your company messaging or commit to initiatives that promise to bring more women into the industry. You need to take a hard look at the culture of your company. Ask yourself, how are women treated? Do they receive the same opportunities as men? Are women subject to harassment and, if so, how are these issues resolved? Are they punished for it? Get out on the job sites. Talk to your lowest-level employees. **LISTEN.**

Male allies, you have no idea the critical role that you play in making this world a better place. My intention is not to paint men as the enemy. The same goes to companies. There are good

---

39 Million Women in Construction Initiative," U.S. Department of Commerce, commerce.gov, accessed August 12, 2024, https://www.commerce.gov/issues/million-women-construction-initiative.

apples and there are bad. There are incredible male souls who've gone to bat for me in my life. So see the injustices. Speak up and have our backs. We thank you. We need you. Keep going to bat for us. Stand up against harassment and discrimination. Advocate for your female colleagues and support their career advancement. And continue to educate yourselves on the challenges women face.

And to our fellow ambitious women, we were born into a society of systemic racism and deep misogyny. We're not going to solve this overnight, but small and consistent actions add up. Keep pushing, keep speaking out, and keep challenging the status quo. Support one another and build networks of solidarity. Demand the respect and opportunities you deserve. Remember, you cannot control how anyone thinks or feels about you. But you can control how you feel, how you show up. Don't let the negative comments or actions of others, triggered more often than not from their own insecurities, control your future. It's your job to grow toward the light.

THE SKILLED PROJECT

The Skilled Project is on a mission to change the perception of the skilled trades, infusing them with dignity, opportunity, power, and fulfillment to mobilize the next generation into these careers. This organization sees storytelling as an immediate first lever to inspire and shine light on the skilled labor jobs of today while highlighting the critical role they play in decarbonizing our built environment and the net new skilled jobs arising from the clean energy transition.

THE
SKILLED
PROJECT

The Skilled Project strives to become an essential resource for people exploring a career in the trades, a partner in reshaping education programs to create more pathways, and an advocate for diverse representation and inclusion across these industries.

Want to get involved or learn more? Reach out to Amanda Luchetti, founder and executive director of the Skilled Project.

Website: https://www.theskilledproject.com/
Instagram: @theskilledproject

PIVOT HIGHER + (HIREHER)

Pivot HIGHer + (Hire Her) is a collective of individuals who support one another through all phases of life and career. A hub of insight, support, and guidance for women and nonbinary individuals. We want to work with you! Pivot Higher works with individuals and organizations to create a community for green and skilled trade jobs.

Feeling stuck? Let us help you navigate today's challenges and beyond. We provide comprehensive and guided support through your questions.

We are here to serve, and help you open the door to your success.

Find us online at https://www.pivothigher.coach/
LinkedIn: LinkedIn.com/company/pivot-higher/
Instagram: pivothigher
TikTok: @pivothigher

Interested in having me as a guest on your podcast or as a keynote speaker?

Reach out to Jessica.Lynn@pivothigher.onmicrosft.com or on LinkedIn at Linked.com/Jlynn89.

I would be delighted to cover a range of topics, including:
- Pathways to skilled trades
- Women in business
- Consulting
- Workforce development
- Inclusive company culture and workplace hiring practices
- Discovering talent for the clean energy transition
- Professional development

And so much more.

REFERENCES

Chapter 1 Start
1 MIG, Metal Inert Gas *Mig Welding: The basics for mild steel*. Miller Electric. (2023, May 26). https://www.millerwelds.com/resources/article-library/mig-welding-the-basics-for-mild-steel

2 Tungsten Inert Gas. Pfaller, A. (2021, September 1). *How a TIG welder works and when to Tig Weld*. Miller Electric. https://www.millerwelds.com/resources/article- library/tig-it-how-a-tig-welder-works-and-when-to-tig-weld#:~:text=TIG%20stands%20for%20tungsten%20inert,the%20tungsten%20and%20weld%20puddle.

Chapter 2 Welder
3 Galvanized steel and zinc poisoning. *Avoid galvanize poisoning when welding*. Red Steel Manufacturing. (2023, May 2). https://www.redsteelmh.com/avoid-galvanize-poisoning-when-welding/

4 Bodywork is the repair of the overall appearance of a car using **different methods, such as fabrication to reshape metal and Bondo to fill** in minor dents. Or replacing entire panels and other necessary repairs all the way up to the paint stage.

5 Metal contraption that lays out the pieces and parts in such a way as to take the guesswork out of assembly. Usually very heavy with clamps.

6 Unhealthy air quality typically occurs from the lack of ventilation, causing the individual to inhale harmful metal fumes and gas by-product - Controlling hazardous fume and gases during welding. (n.d.). https://www.osha.gov/sites/default/files/publications/OSHA_FS-3647_Welding.pdf

7 Plastic welding is the process of melting plastic into a mold or joining two pieces of plastic by melting the surface. Usually uses a mix of pressure and heat.

Chapter 3 Construction Management
8 Liquidated damages are when a penalty of a high amount, sometimes in the tens of thousands of dollars, is imposed on a construction for not being completed on time.

9 Juris Doctor, or Doctor of Law, is a postgraduate diploma upon the completion of all education required to practice law.

Chapter 4 California
10 Dinen, C. (2018, September 4). *Why being scared to relocate isn't such a bad thing.* https://chelseadinen.com/scared-to-relocate/

11 Nomex is a bright-orange jumpsuit made of fire-retardant material that prevents the individual from going up in a blaze. This is essential personal protective equipment in a refinery.

12 Post-weld heat treatment occurs after a pipe has been welded to help with the expansion of the gas or fluid line. Typically, little things that look like pink chiclets are wrapped around the pipe and turn red when hooked up to an electrical current.

REFERENCES

13 Preheat weld, similar to post-weld, occurs prior to the pipe being welded. Chiclets are set up around a weld or seam opening, then heated up when the welder fuses the gap together.

14 A bomb box is, in essence, a large stack of what appears to be a shipping container. It protects people during a refinery explosion.

15 The pipe fitter preps, installs, and repairs a gas or fluid line. Apprentices will fit up or line the two pieces of pipe while the welder fuses the seam or gap closed.

16 Headache shack, or mobile digital plan station *Jobsite Boxes*. KNAACK. (n.d.). https://www.knaack.com/products/jobsite-storage-solution/jobsite-boxes/FieldStationSeries/118-01

17 PPE stands for personal protective equipment. It typically includes safety eyewear, gloves, a vest or Nomex, and steel-toe or composite boots. These are the basics to keep yourself safe on a construction site.

18 Heavy-duty masks can range based on function. In this instance, they were a full-face covering with two air filters that stuck out. Think postapocalyptic movie, and you'll get the idea.

19 TEAMs are a specific subcontractor focused on managing different types of data. At the refinery, there were two dedicated individuals who monitored the electrical feedback from the post-weld and preheat electrical lines when they were turned on and cooking a pipe. *Integrity management: Asset integrity for pipeline & aviation*. TEAM, Inc. (2024, May 21). https://www.teaminc.com/

20 A PIP is typically outlined to manage a specific focus of professional development. Also used as a last resort to document poor behavior or an underperforming employee before they are fired.

Chapter 5 Denver
21 Laborers are the unsung heroes of any job site. They run the gambit of work performed, including picking up trash, cleaning, and unloading or loading tools or materials.

Chapter 6 Oregon
22 Architect, engineering, and contractor (AEC) meetings are half-day sessions that bring together key decision-makers to review the project process, ask questions, and resolve problems.

23 An RFI (request for information) is a typical communication tool between contractors, engineers, or architects used to ask a question or potentially suggest a change to a scope of work.

24 Procore is industry-standard construction management software. It's used for scheduling, reviewing drawings, and communicating changes to a project. Request a demo. Procore. (n.d.). https://www.procore.com.

25 Blue steel, a specific mix of alloys and external processing, is typically very expensive due to the special treating of steel. *Blueing of Steel.* Custom Machined, Forged, Cast & Plated Parts - Bunty LLC. (2020b, May 29). https://buntyllc.com/blueing/

26 Men will verbally assault or write profane comments, calling women names, such as bitches, hos, etc., on a job site. These words can also be seen written in Sharpie on temporary wall framing or with a finger on a dirty window. In the porta-johns, I would see pictures galore of male

REFERENCES

anatomy. On several occasions, we had reports of someone smearing human feces all over the seat, door, and whatever else they could touch inside a porta-john. This is why the women's porta-johns are often locked. It gives women a hope of having a dedicated space to take care of business.

Chapter 7 Pandemic
27 Milestone work is a set point in a schedule where progress is tracked or compared. As an example, project kickoff and final turnover are project milestones.

28 An NDA, or nondisclosure agreement, is a signed contract or document in which you sign away all rights to take the other party involved to court or sue them for any type of damages, neglect, abuse, etc.

29 COBRA is complicated and expensive post-employment insurance that offers temporary benefits after group coverage ends.

30 Intergovernmental Panel on Climate Change (IPCC) is an incredibly important group of scientists, activists, and environmentalists working for the United Nations. They collect data and share it with key decision-makers across the globe. Check out https://www.ipcc.ch/ for more info.

Chapter 8 MBA
31 Farmers the company was pursuing tended to already be making serious money in the millions of dollars. This was a missed opportunity to reach a wider community of folks, including information about better access to renewables for disenfranchised communities.

32 Agile planning methods are a form of Lean project management. This approach focuses on collaboration and offers flexibility in the schedule. Learn more here at PMI.org.

33 A capstone is used as a final marker or milestone to show off what was learned. It's a performance indicator to embody a subject or overall comprehension.

Chapter 9 Austin
34 The Best Builder team consisted of dedicated experts who evangelized the Lean building process.

Chapter 10 Pivot
35 UX/UI designers create user-centric experiences. They focus on the needs of the target market and building a solution that serves them.

36 Solid oxide fuel cells (SOFCs) offer a specific process of capturing gasses and transforming them with the use of electrical current and cathode materials. Think of lots of ceramic and electrical current in a pretty steel box. Bloom Energy. (2022, May 19). Everything you need to know about solid oxide fuel cells. https://www.bloomenergy.com/blog/everything-you-need-to-know-about-solid-oxide-fuel-cells/

Chapter 11 Director
37 Women in the workforce facts and statistics - Glassdoor career guides. (June 6, 2021). https://www.glassdoor.com/blog/guide/women-in-the-workforce-statistics/

REFERENCES

Afterword The Skilled Project

[38] Liz Elting, "The High Cost Of Ambition: Why Women Are Held Back For Thinking Big," *Forbes.com,* April 24, 2017, https://www.forbes.com/sites/lizelting/2017/04/24/the-high-cost-of-ambition-why-women-are-held-back-for-thinking-big/.

[39] "Million Women in Construction Initiative," U.S. Department of Commerce, commerce.gov, accessed August 12, 2024, https://www.commerce.gov/issues/million-women-construction-initiative.

ABOUT THE AUTHOR

Jessica (Jessi) Lynn is a 2023 MBA graduate from the University of Minnesota's Carlson School of Management. With over eight years of combined managerial experience in the energy and construction industry, she offers robust leadership skills. Her journey has been marked by a passion for innovation and a commitment to making a difference.

In her current role as the director of strategic initiatives for sustainable infrastructure, her mission is clear: amplify and elevate field voices. Each day, she takes an active role in shaping the future of sustainable practices through encouraging inclusive conversations and collaboration, refining processes, and building the right tools for the job.

Jessi is also the visionary behind Pivot HIGHer+ (Hire Her), a social enterprise focused on women and nonbinary who are career changers. Her team focuses on providing resources and support through connecting community and building a holistic resource hub. Pivot HIGHer+ is that bridge—a pathway for women and nonbinary to thrive in new roles, break down barriers, and empower change.

When she's not immersed in workforce development and community endeavors, Jessi spends time traveling with friends, and her pupper, Suki. Jessi is a force to be reckoned with—a trailblazer who's shaping a more equitable and sustainable future, one conversation and one initiative at a time.

You can connect with her on:
https://www.linkedin.com/in/jlynn89
https://www.instagram.com/truth_woman_in_construction

Subscribe to her newsletter, *Sustainability Call to Action:*
https://www.linkedin.com/build-relation/newsletter-follow?entityUrn=7069371636109508608

Made in the USA
Monee, IL
25 September 2024

66577887R00138